Lecture Notes in Physics

New Series m: Monographs

The Editorial Policy for Monographs

The series Lecture Notes in Physics reports new developments in physical research and teaching - quickly, informally, and at a high level. The type of material considered for publication in the New Series m includes monographs presenting original research or new angles in a classical field. The timeliness of a manuscript is more important than its form, which may be preliminary or tentative. Manuscripts should be reasonably self-contained. They will often present not only results of the author(s) but also related work by other people and will provide sufficient motivation, examples, and applications.
The manuscripts or a detailed description thereof should be submitted either to one of the series editors or to the managing editor. The proposal is then carefully refereed. A final decision concerning publication can often only be made on the basis of the complete manuscript, but otherwise the editors will try to make a preliminary decision as definite as they can on the basis of the available information.
Manuscripts should be no less than 100 and preferably no more than 400 pages in length. Final manuscripts should preferably be in English, or possibly in French or German. They should include a table of contents and an informative introduction accessible also to readers not particularly familiar with the topic treated. Authors are free to use the material in other publications. However, if extensive use is made elsewhere, the publisher should be informed. Authors receive jointly 50 complimentary copies of their book. They are entitled to purchase further copies of their book at a reduced rate. As a rule no reprints of individual contributions can be supplied. No royalty is paid on Lecture Notes in Physics volumes. Commitment to publish is made by letter of interest rather than by signing a formal contract. Springer-Verlag secures the copyright for each volume.

The Production Process

The books are hardbound, and quality paper appropriate to the needs of the author(s) is used. Publication time is about ten weeks. More than twenty years of experience guarantee authors the best possible service. To reach the goal of rapid publication at a low price the technique of photographic reproduction from a camera-ready manuscript was chosen. This process shifts the main responsibility for the technical quality considerably from the publisher to the author. We therefore urge all authors to observe very carefully our guidelines for the preparation of camera-ready manuscripts, which we will supply on request. This applies especially to the quality of figures and halftones submitted for publication. Figures should be submitted as originals or glossy prints, as very often Xerox copies are not suitable for reproduction. In addition, it might be useful to look at some of the volumes already published or, especially if some atypical text is planned, to write to the Physics Editorial Department of Springer-Verlag direct. This avoids mistakes and time-consuming correspondence during the production period.
As a special service, we offer free of charge LaTeX and TeX macro packages to format the text according to Springer-Verlag's quality requirements. We strongly recommend authors to make use of this offer, as the result will be a book of considerably improved technical quality. The typescript will be reduced in size (75% of the original). Therefore, for example, any writing within figures should not be smaller than 2.5 mm.
Manuscripts not meeting the technical standard of the series will have to be returned for improvement.
For further information please contact Springer-Verlag, Physics Editorial Department II, Tiergartenstrasse 17, D-69121 Heidelberg, FRG.

Gennady P. Berman Evgeny N. Bulgakov
Darryl D. Holm

Crossover-Time in Quantum Boson and Spin Systems

Springer-Verlag
Berlin Heidelberg New York
London Paris Tokyo
Hong Kong Barcelona
Budapest

Authors

Gennady P. Berman
Center for Nonlinear Studies
Los Alamos National Laboratory, MS B258
Los Alamos, NM 87545, USA
email: gpb@goshawk.lanl.gov
and
Kirensky Institute of Physics
660036, Krasnoyarsk, Russia
email: tnp@iph.krasnoyarsk.su

Evgeny N. Bulgakov
Kirensky Institute of Physics
660036, Krasnoyarsk, Russia
email: tnp@iph.krasnoyarsk.su

Darryl D. Holm
Theoretical Division
Los Alamos National Laboratory, MS B284
Los Alamos, NM 87545, USA
email: dholm@lanl.gov

ISBN 3-540-58011-5 Springer-Verlag Berlin Heidelberg New York
ISBN 0-387-58011-5 Springer-Verlag New York Berlin Heidelberg

CIP data applied for.

Printed in Germany

Printing and binding: Druckhaus Beltz, Hemsbach/Bergstr.
Typesetting: camera-ready by author/editor
SPIN: 10080256 55/3140-543210 - Printed on acid-free paper

Acknowledgements

Two of the authors (GPB and ENB) would like to thank D. Cohen, G. Doolen and J.M. Hyman of The Center for Nonlinear Studies, Los Alamos National Laboratory, for their hospitality during the time when this work was perfomed. GPB would also like to thank K.N. Alekseev, A.M. Iomin, F.M. Izrailev, A.R. Kolovsky, V.I. Tsifrinovich, G.M. Zaslavsky, with whom many of the results discussed in this review were derived, and J. Bayfield, B.V. Chirikov, P. Constantin, J. Lebowitz, T. Seligman, V.E. Zakharov for useful discussions and remarks. We are also grateful to J.E. Marsden for encouragement and helpful suggestions. This work was partially supported by funding from U.S. DOE, Office of Scientific Computation. GPB and ENB also gratefully acknowledge the partial support of the Linkage Grant 93-1602 from the NATO Special Programme Panel on Nanotechnology. Finally, we are each deeply grateful to our families for their welcome distractions during our writing.

Contents

1 Introduction

We consider quantum dynamical systems (in general, these could be either Hamiltonian or dissipative, but in this review we shall be interested only in quantum Hamiltonian systems) that have, at least formally, a classical limit. This means, in particular, that each time-dependent quantum-mechanical expectation value $X_i(t)$ has as $\hbar \to 0$ a limit $X_i(t) \to X_i^{(cl)}(t)$ of the corresponding classical system. Quantum-mechanical considerations include an additional dimensionless parameter $\epsilon = \hbar/const.$ connected with the Planck constant $\hbar$. Even in the quasiclassical region where $\epsilon \ll 1$, the dynamics of the quantum and classical functions $X_i(t)$ and $X_i^{(cl)}(t)$ will be different, in general, and quantum dynamics for expectation values may coincide with classical dynamics only for some finite time. This characteristic time-scale, $\tau_\hbar$, could depend on several factors which will be discussed below, including: choice of expectation values, initial state, physical parameters and so on. Thus, the problem arises in this connection: How to estimate the characteristic time-scale $\tau_\hbar$ of the validity of the quasiclassical approximation and how to measure it in an experiment ? For rather simple integrable quantum systems in the stable regions of motion of their corresponding classical phase space, this time-scale $\tau_\hbar$ usually is of order (see, for example, [2])

$$\tau_\hbar = \frac{const}{\bar{\mu}\hbar^\alpha}, \tag{1.1}$$

where $\bar{\mu}$ is the dimensionless parameter of nonlinearity (discussed below) and α is a constant of the order of unity.

Recently this problem has attracted additional interest in connection with the investigation of dynamical and spectral properties of nonintegrable quantum systems (the so-called problem of "quantum chaos") [3-15]. In this case, the classical dynamical "zero approximation" appears to be strongly unstable and chaotic. So, the appearance of quantum effects even in the quasiclassical region of parameters may occur significantly sooner than (1.1). For example, in certain simple quantum dynamical systems with 1.5 degrees of freedom, (see also below in this review) the time-scale $\tau_\hbar$ is of order

[1] (see also [3-12])

$$\tau_\hbar = C_1 \ell n(C_2/\hbar), \tag{1.2}$$

(where $C_{1,2}$ are constants and C_1^{-1} is proportional to the Lyapunov exponent). This means that for times $\tau < \tau_\hbar$ quantum dynamics differs only slightly from a classical chaotic motion, but for times $\tau > \tau_\hbar$ quantum corrections play a significant role. So, at times $\tau \sim \tau_\hbar$ the character of dynamics of such a quantum system undergoes qualitative modification. Then, the problem arises of the possibility of an experimental verification of the time-scale (1.2), the time at which quantum and classical dynamics first begin to differ. This problem is rather important as the logarithmic time scale $\tau_\hbar$ is believed to be the smallest time at which quantum and classical dynamics differ measurably. To succeed in this measurement, one must identify an experiment involving a proper quantum system in its quasiclassical region of parameters for which the time-scale $\tau_\hbar$ could be observed. In this connection note that the experiments recently carried out [16,17] with Rydberg states of a hydrogen atom in a resonant microwave field and in the region of parameters for quantum chaos allow the study of processes like diffusion of an electron up to the ionization threshhold, as well as multiphoton processes, etc. [16-19]. Nevertheless, even for the initial population of an electron in the region with a principal quantum number $n \sim 50 - 100$ this Rydberg system can not be considered as sufficiently quasiclassical to observe the time scale $\tau_\hbar$ (1.2), as in this case the time-scale $\tau_\hbar$ is too small for experimental observation. Note that in the region of a *regular* classical dynamics one *could* perhaps observe experimentally in this system the time-scale $\tau_\hbar$ (1.1), which in this case has an order $\tau_\hbar = \tau\omega \sim 10$, where ω is a characteristic microwave frequency and $n \sim 100$. In this review we consider the kind of systems in which the transition from a classical Hamiltonian chaos to an essentially quantum dynamics could be observed experimentally [20-23]. One of these systems consists of an ensemble of N two-level atoms in a resonator interacting with a single resonant eigenmode. We assume that the atoms are interacting with an additional external coherent field with a frequency slightly different from the frequency of atomic transition [22]. Also, we consider a system of N quantum paramagnetic spins in a resonator interacting with a single resonant eigenmode and with a magnetic field which includes a constant component and a periodically time-dependent component [23]. All these systems [20-23] are nonintegrable generalizations of the well known integrable Dicke model [24] from nonlinear optics. The Dicke model is completely integrable, either classically or quantum-mechanically. In contrast, we show that the above-mentioned systems exhibit semiclassical dynamics (when the radiation field is considered classically) that can make a transition to chaos at parameter values which are rather common in experiments, for example, in the recent experiments on Rydberg atoms in a microwave cavity [25-27]. In these systems, the quasiclassical parameter ϵ is of order: $\epsilon = 1/N << 1$, where N can reach the value $N \sim 10^{10}$ in the nonlinear optical systems and $N \sim 10^{22}$ in the paramagnetic spin system. So, the systems we analyze here can be prepared in a "deep quasiclassical region"($N \gg 1$), and possibly the time-scale $\tau_\hbar$ (1.2) could be observed in this case in the nondissipative regime. It is shown, that for these systems $\tau_\hbar \sim \ln N$, and quantum correlation functions grow exponentially for the case of unstable and chaotic classical motion. One of our aims in this review is to attract the attention of experimentalists to measuring the time-scale $\tau_\hbar$ and quantum correlation functions in such deeply quasiclassical systems.

We give in this review a consequent analysis of the different peculiarities of the time-scale $\tau_\hbar$ for integrable and nonintegrable quantum systems starting from its definition and including a consideration of quantum dynamics in boson and spin coherent states, in Wigner representation and others. We consider the cases of either strong or weak quantum chaos. To analyze the difference between quantum and classical dynamics in a "deep quasiclassical region" we mainly use a closed set of c-number equations derived here for quantum expectation values and for quantum correlation functions. Then, we use an additional averaging over the initial density matrix.

Also at the end of this review we present results on the quantum dynamics of nonlinear Hamiltonians in the so-called Stationary Coherent States [28,29]. Although these quantum Hamiltonians look rather artificial, the time-dependent expectation values in these states realize the usual classical behaviour and $\tau_\hbar = \infty$ in this case. We show that in these states quantum Hamiltonian dynamics can exhibit a classical chaos even of a strange attractor type (!). We decided to review the results on the quantum dynamics in these states as well, because these states may be interesting in connection with a construction of a bridge between quantum and classical approaches for "macroscopic classical systems". We also present here the proofs of some of the most frequently used expressions.

2 Method of Coherent States

In this review we are mainly interested in comparing quantum and classical dynamics for several integrable and nonintegrable boson and spin systems. A convenient basis for this kind of approach is provided by the so-called coherent states (CS)[30-37]. In this section we present some notations for boson and spin CSs which will be widely used in the sections of that follow. In addition, we review the standard properties of boson and spin CSs, and then collect some identities and expressions that will be useful in discussing exact quantum dynamics of expectation values.

2.1 Introduction to the Coherent States

In quantum mechanics, measurement of a real quantity corresponds to determining the expectation value of some Hermitian operator, **A**. The term Hermitian means

$$\mathbf{A}^+ = \mathbf{A}, \tag{2.1}$$

where $\mathbf{A}^+$ is the operator Hermitian adjoint to the operator $\mathbf{A}$ (complex-conjugate transpose for matrices). The following condition is fulfilled

$$< \Psi|\mathbf{A}^+|\Phi >= (< \Phi|\mathbf{A}|\Psi >)^*, \tag{2.2}$$

where $|\Psi >$ and $|\Phi >$ are wave functions (or states) written in Dirac notation. Wave functions chosen in the different representations depend on the corresponding internal variables, and the operation $< ... >$ means the integration or summation over all the independent variables. For example, in x-representation we have matrix element

$$< \Psi|\mathbf{A}|\Phi >\equiv \int \Psi^*(x)\mathbf{A}\Phi(x)dx.$$

By using an explicit form of the operator $\mathbf{A}$ and the state $|\Psi>$, one can define the expectation value of the operator $\mathbf{A}$ in the state $|\Psi>$ and its standard deviation (variance) σ_A:

$$A =< \Psi|\mathbf{A}|\Psi >, \qquad \sigma_A^2 =< \Psi|(\mathbf{A} - A)^2|\Psi > . \tag{2.3}$$

It is clear that the expectation value A and the variance σ_A both depend on the state $|\Psi>$ in which the system is prepared. (In different states $|\Psi>$, one will obtain different values for A and σ_A).

In what follows, we shall need the quantum states $|\Psi>$ which are closest to the classical ones in a certain sense. One choice of such states is based on minimizing the well known Heisenberg uncertainty relation,

$$\sigma_A\sigma_B \quad \geq \quad \frac{1}{2}| < \Psi|[\mathbf{A},\mathbf{B}]|\Psi > |. \tag{2.4}$$

In (2.4) the operators $\mathbf{A}$ and $\mathbf{B}$ are two Hermitian operators with commutator

$$[\mathbf{A},\mathbf{B}] \equiv \mathbf{AB} - \mathbf{BA}. \tag{2.5}$$

We recall here a useful and rather general proof of the inequality (2.4) (which does not depend on the choice of system) (see [38]). For this, define a state

$$|\Phi >= \mathbf{F}|\Psi >, \tag{2.6}$$

where $\mathbf{F}$ is the non-Hermitian operator given by the linear combination

$$\mathbf{F} = \mathbf{A} + i\lambda\mathbf{B} \tag{2.7}$$

and λ is a real number. Then, from positivity of the norm of the state $|\Phi>$ in (2.6), we have the following relation

$$\parallel \Phi \parallel^2 =< \Psi|(\mathbf{A} - i\lambda\mathbf{B})(\mathbf{A} + i\lambda\mathbf{B})|\Psi > \tag{2.8}$$

$$= < \mathbf{A}^2 > + \lambda^2 < \mathbf{B}^2 > + i\lambda < [\mathbf{A}, \mathbf{B}] > \quad \geq 0.$$

In (2.8) we introduce the notation

$$< \mathbf{D} > \equiv < \Psi|\mathbf{D}|\Psi > \equiv D.$$

From (2.8) it follows: a) $< [\mathbf{A}, \mathbf{B}] >$ is an imaginary value; and b) the right part in (2.8) has a minimum at the value

$$\lambda_0 = -\frac{i}{2} < [\mathbf{A}, \mathbf{B}] > / < \mathbf{B}^2 > . \tag{2.9}$$

From (2.8) and (2.9) one has for the norm of the state $|\Phi >$ at the point λ_0 (2.9)

$$\| \Phi \|^2_{\lambda_0} = < \mathbf{A}^2 > + \frac{1}{4}(< [\mathbf{A}, \mathbf{B}] >)^2 / < \mathbf{B}^2 > \quad \geq 0. \tag{2.10}$$

Then, it follows from (2.10)

$$< \mathbf{A}^2 >< \mathbf{B}^2 > \quad \geq \quad -\frac{1}{4}(< [\mathbf{A}, \mathbf{B}] >)^2. \tag{2.11}$$

As the operators $\mathbf{A}$ and $\mathbf{B}$ are arbitrary Hermitian operators we can choose instead of them the following ones

$$\mathbf{A} \to \mathbf{A} - A, \qquad \mathbf{B} \to \mathbf{B} - B$$

with the same commutator

$$[\mathbf{A} - A, \mathbf{B} - B] = [\mathbf{A}, \mathbf{B}]. \tag{2.12}$$

Thus, using the condition that $< [\mathbf{A}, \mathbf{B}] >$ is imaginary leads from (2.11) to

$$< (\mathbf{A} - A)^2 >< (\mathbf{B} - B)^2 > \quad \geq \quad \frac{1}{4}| < [\mathbf{A}, \mathbf{B}] > |^2, \tag{2.13}$$

which proves the inequality (2.4).

We apply now the inequality (2.4) to a canonically conjugate pair of operators, for which $[\mathbf{A}, \mathbf{B}] = i\hbar$. In this case we have from (2.13) the well known Heisenberg uncertainty condition

$$\sigma_A \sigma_B \geq \frac{\hbar}{2}. \tag{2.14}$$

In particular, for the canonically conjugate coordinate and momentum

$$\mathbf{A} = \mathbf{x}, \qquad \mathbf{B} = \mathbf{p}$$

we have from (2.14)

$$\sigma_x \sigma_p \geq \frac{\hbar}{2}. \tag{2.15}$$

Let find a state $|\Psi>$ in which the product of uncertainties of a coordinate and a momentum in (2.15) is minimal. Additionally we shall require a normalization condition for the state $|\Psi>$, i.e. fulfillment of the condition

$$< \Psi|\Psi >= 1. \tag{2.16}$$

Then, the problem consists in minimizing of the functional expression for $\sigma_x \sigma_p$ with the additional condition (2.16) [32]. Introduce the functional

$$U =< \Psi|(\mathbf{x} - x)^2|\Psi >< \Psi|(\mathbf{p} - p)^2|\Psi > \tag{2.17}$$

$$+\nu(< \Psi|\Psi > -1),$$

where

$$x \equiv < \mathbf{x} > \equiv < \Psi|\mathbf{x}|\Psi >, \qquad p \equiv < \mathbf{p} > \equiv < \Psi|\mathbf{p}|\Psi >,$$

and ν is a Lagrange multiplier. Then, the minimization problem reduces to

$$\frac{\delta U}{\delta < \Psi|} = \left[\frac{(\mathbf{x} - x)^2}{\sigma_x^2} + \frac{(\mathbf{p} - p)^2}{\sigma_p^2} - 2 \right] |\Psi >= 0. \tag{2.18}$$

In the x-representation the operator $\mathbf{p}$ has the form $\mathbf{p} =$ -$i\hbar\partial/\partial x$ and it is easy to check that the normalized solution of the differential equation (2.18) has the form

$$|\Psi(x) >= (2\pi\sigma_x^2)^{-1/4} \tag{2.19}$$

$$\times \exp\left(-\frac{(x- < \mathbf{x} >)^2}{4\sigma_x^2} + i\frac{< \mathbf{p} > (x- < \mathbf{x} >)}{\hbar} + i\frac{< \mathbf{x} >< \mathbf{p} >}{2\hbar} \right).$$

In the state (2.19) the following condition is fulfilled

$$\sigma_x \sigma_p = \frac{\hbar}{2}, \tag{2.20}$$

for an arbitrary σ_x (the constant phase in (2.19) is introduced for convenience, see below). The state (2.19) provides a minimum of the uncertainty condition (2.15) and is called a coherent state (CS) [30]. The CS (2.19) and its variants and generalizations (see, for example, [34,35]) are of considerable importance in several fields of physics, largely because the expectation values ($<\mathbf{A}>$) and the standard deviations (σ_A) are so well defined in these states.

The CSs are also important because they (or states close to them) are often realized in experiments, especially in the quasiclassical region of parameters (see, for example, [32,39]). A great number of scientific papers are devoted to the description of the various properties of the CSs (see [30-37] and references therein). Next we present a few properties of CSs that will be used in the following.

2.2 Some Properties of Boson Coherent States

From the considerations given above it is clear that the CS $|\Psi>$ in (2.19) needs not be connected directly with any specific quantum-mechanical system. At the same time it is very useful to consider CSs from the point of view of a simple quantum-mechanical system, namely, the one-dimensional quantum linear oscillator, whose Hamiltonian is [32]

$$\mathbf{H} = \frac{\mathbf{p}^2}{2m} + \frac{m\omega^2\mathbf{x}^2}{2}. \tag{2.21}$$

Instead of canonically conjugate operators $\mathbf{x}$ and $\mathbf{p}$ we introduce in the usual way two other canonically conjugate operators: a creation operator $\mathbf{a}^+$ and an annihilation operator $\mathbf{a}$, given by

$$\mathbf{a} = \sqrt{\frac{m\omega}{2\hbar}}(\mathbf{x} + i\mathbf{p}/m\omega), \tag{2.22}$$

$$\mathbf{a}^+ = \sqrt{\frac{m\omega}{2\hbar}}(\mathbf{x} - i\mathbf{p}/m\omega).$$

These operators satisfy the commutation relation

$$[\mathbf{a}, \mathbf{a}^{+}] = 1. \tag{2.23}$$

It is known [30] that the eigenfunction $|\alpha>$ of the operator $\mathbf{a}$ in (2.22) is a CS as in (2.19). In fact, the eigen-problem for the operator $\mathbf{a}$

$$\mathbf{a}|\alpha>= \alpha|\alpha> \tag{2.24}$$

has in the x-representation a normalized solution

$$|\alpha>= \left(\frac{m\omega}{\pi\hbar}\right)^{1/4} \exp\left[-\left(\sqrt{\frac{m\omega}{2\hbar}}x - \alpha\right)^2 + \frac{\alpha^2}{2} - \frac{|\alpha|^2}{2}\right], \tag{2.25}$$

where α is an arbitrary complex number. It is easy to see from (2.22) and (2.24) that in CS (2.25) the following relations are satisfied

$$\sigma_x^2 =< \alpha|(\mathbf{x} - x)^2|\alpha >= \hbar/2m\omega, \tag{2.26}$$

$$\sigma_p^2 =< \alpha|(\mathbf{p} - p)^2|\alpha >= \hbar m\omega/2,$$

$$\alpha = \sqrt{\frac{m\omega}{2\hbar}}(x + ip/m\omega).$$

Substituting (2.26) in (2.25) gives the expression $|\Psi(x)>$ in (2.19). The CS (2.25) still has no connection with the Hamiltonian (2.21). However, now we expand the CS (2.25) with an arbitrary complex value α in a complete set of eigenfunctions of the operator $\mathbf{H}$ in (2.21) [32]. For this, we use the well known solutions $u_n(x)$ of the eigenvalue problem for the operator $\mathbf{H}$ (2.21)

$$\mathbf{H}u_n(x) = E_n u_n(x), \tag{2.27}$$

where the eigenfunction $u_n(x)$ and the eigenvalue E_n have the form

$$u_n(x) \equiv |n>= \left(\frac{1}{2^n n!}\sqrt{\frac{m\omega}{\pi\hbar}}\right)^{1/2} H_n\left(\sqrt{\frac{m\omega}{\hbar}}x\right) \exp\left(-\frac{m\omega x^2}{2\hbar}\right), \tag{2.28}$$

$$E_n = \hbar\omega\left(n + \frac{1}{2}\right), \quad n = 0, 1, 2, ...$$

In (2.28) $H_n(z)$ is a Hermite polynomial. Also we recall the expression for the generating function for Hermite polynomials

$$\sum_{n=0}^{\infty} \frac{\xi^n}{n!} \exp(-x^2) H_n(x) = \exp\left[-(x-\xi)^2\right]. \tag{2.29}$$

Using (2.28), (2.29) expresses the CS $|\alpha>$ in (2.25) in the form

$$|\alpha>= e^{-\frac{|\alpha|^2}{2}} \sum_{n=0}^{\infty} \frac{\alpha^n}{\sqrt{n!}} |n> . \tag{2.30}$$

Remarks [30-33]:

1). In the CS $|\alpha>$ the uncertainty is minimal: $\sigma_x \sigma_p = \hbar/2$.

2). The CS $|\alpha>$ is normalized: $<\alpha|\alpha>= 1$.

3). The physical meaning of the complex value α is the following. The probability for n quanta to be in the state $|\alpha>$ is given by

$$P_n(\alpha) \equiv |<n|\alpha>|^2 = \frac{e^{-|\alpha|^2}|\alpha|^n}{n!} = \frac{e^{-\bar{n}}\bar{n}^n}{n!}, \tag{2.31}$$

where

$$\bar{n} \equiv |\alpha|^2.$$

The distribution (2.31) is the Poisson distribution with the average value of quanta $\bar{n} = |\alpha|^2$. By increasing $|\alpha|$ it is possible to construct a CS $|\alpha>$ in "deep" quasiclassical region ($\bar{n} \gg 1$).

4). CSs with different α are not orthogonal, since

$$<\alpha|\beta>= \exp\left(-\frac{|\alpha|^2}{2} - \frac{|\beta|^2}{2} + \alpha^*\beta\right). \tag{2.32}$$

5). The CS $|\alpha>$ can be constructed using a shift operator $\mathbf{D}_s(\alpha, \alpha^*)$

$$\mathbf{D}_s(\alpha, \alpha^*) = exp(\alpha \mathbf{a}^+ - \alpha^* \mathbf{a}), \tag{2.33}$$

$$|\alpha>= \mathbf{D}_s(\alpha, \alpha^*)|0>, \tag{2.34}$$

where $|0>$ is the CS with $\alpha = 0$ and at the same time is the eigenstate of the Hamiltonian $\mathbf{H}$ (2.21) with $n = 0$ (the vacuum state with the energy $E_0 = \hbar\omega/2$). To prove expression (2.34), it is possible to use the well known relations for the creation and annihilation operators

$$\mathbf{a}|n>=\sqrt{n}|n-1>, \qquad (\mathbf{a}|0>=0), \tag{2.35}$$
$$\mathbf{a}^+|n>=\sqrt{n+1}|n+1>,$$
$$\mathbf{a}^+\mathbf{a}|n>=n|n>,$$
$$|n>=(1/\sqrt{n!})(\mathbf{a}^+)^n|0>,$$

and the following expression

$$\mathbf{D}_s(\alpha,\alpha^*)=e^{\frac{|\alpha|^2}{2}}e^{\alpha\mathbf{a}^+}e^{-\alpha^*\mathbf{a}}. \tag{2.36}$$

The expression (2.36) follows from a more general relation for two arbitrary operators **A** and **B** whose commutator [**A**, **B**] is a c-number [33]

$$e^{\mathbf{A}+\mathbf{B}}=e^{-\frac{1}{2}[\mathbf{A},\mathbf{B}]}e^{\mathbf{A}}e^{\mathbf{B}}=e^{-\frac{1}{2}[\mathbf{B},\mathbf{A}]}e^{\mathbf{B}}e^{\mathbf{A}}. \tag{2.37}$$

In our case (2.33) $\mathbf{A}=\alpha\mathbf{a}^+$, $\mathbf{B}=-\alpha^*\mathbf{a}$.

6). The CSs form a complete system [30-32]

$$\frac{1}{\pi}\int d^2\alpha|\alpha><\alpha|=1, \qquad (d^2\alpha\equiv d(Re\alpha)d(Im\alpha)).$$

7). One of the interesting properties of the CSs is the conservation of coherence of the wave packet for the system with the Hamiltonian (2.21). Let, for example, the system with the Hamiltonian (2.21) be initially at time t=0 in the CS $|\alpha_0>$, i.e.,

$$|\Psi(t=0)>=|\alpha_0>. \tag{2.38}$$

Then, the solution of the Schrödinger equation

$$i\hbar\frac{\partial|\Psi(t)>}{\partial t}=\mathbf{H}|\Psi(t)> \tag{2.39}$$

at an arbitrary time t is given by

$$|\Psi(t)>=e^{-\frac{i}{\hbar}\mathbf{H}t}|\alpha_0>=e^{-i\omega(\mathbf{a}^+\mathbf{a}+1/2)t}|\alpha_0>. \tag{2.40}$$

Then, using equations (2.30) and (2.35) with (2.40) gives

$$|\Psi(t)> = e^{-\frac{i\omega t}{2}} e^{-\frac{|\alpha_0|^2}{2}} \sum_{n=0}^{\infty} \frac{[\alpha_0 \exp(-i\omega t)]^n}{\sqrt{n!}} |n> \tag{2.41}$$

$$= e^{-\frac{i\omega t}{2}} |\alpha_0 \exp(-i\omega t) > \equiv e^{-i\frac{\omega t}{2}} |\alpha(t)> .$$

Equation (2.41) shows that the initial CS $|\alpha_0>$ remains a CS for an arbitrary time t, but the value α_0 changes according to

$$\alpha(t) = \alpha_0 \exp(-i\omega t). \tag{2.42}$$

Using (2.26) and (2.42) gives the time dependent expectation values $x(t)$ and $p(t)$

$$x(t) = x_0 \cos(\omega t) + \frac{p_0}{m\omega} \sin(\omega t), \tag{2.43}$$

$$p(t) = p_0 \cos(\omega t) - m\omega x_0 \sin(\omega t),$$

which formally coincide with the classical solutions. It is well known that such coincidence of quantum and classical dynamics for expectation values occurs not only for the Hamiltonian (2.21), but for any quadratic Hamiltonian of the type,

$$\mathbf{H} = A(t)\mathbf{a}^+ + A^*(t)\mathbf{a}^+ + B(t)\mathbf{a}^+\mathbf{a} + C(t)(\mathbf{a}^+)^2 + C^*(t)\mathbf{a}^2, \tag{2.44}$$

(where $A(t), B(t), C(t)$ are arbitrary time-dependent c-number functions), and for an arbitrary initial state $|\Psi(0)>$. This is because the Hamiltonian (2.44) gives the Heisenberg equation, for example, for operator **a**

$$i\hbar\dot{\mathbf{a}} = [\mathbf{a}, \mathbf{H}] \tag{2.45}$$

which is linear and the time-dependent expectation value in this case actually coincides with its classical limit. The CS is special in this regard: if we were to start at t=0 with an arbitrary $|\Psi(0)>$ different from CS, the evolution would not conserve the coherent properties of the initial wave packet, even for quadratic Hamiltonians of the type (2.44).

8). Finally we collect a few other properties of boson CSs that will be used in the following to derive closed exact c-number equations for time-dependent expectation values of rather general quantum Hamiltonians. Consider an arbitrary function $\mathbf{f}$ of boson operators that possesses a formal power series expansion

$$\mathbf{f} = \mathbf{f}(\mathbf{a}^+, \mathbf{a}). \tag{2.46}$$

Then, the previous remarks make it easy to prove the following formulas [31] (see also Appendix A)

$$< z|\mathbf{f}\mathbf{a}^+|z >= e^{-|z|^2} \frac{\partial}{\partial z} e^{|z|^2} f, \tag{2.47}$$

$$< z|\mathbf{a}^+\mathbf{f}|z >= z^* f, \qquad < z|\mathbf{f}\mathbf{a}|z >= z f,$$

$$< z|\mathbf{a}\mathbf{f}|z >= e^{-|z|^2} \frac{\partial}{\partial z^*} e^{|z|^2} f,$$

where f is the expectation value of the operator $\mathbf{f}$ in CS $|\alpha >$, namely

$$f =< \alpha|\mathbf{f}|\alpha > . \tag{2.48}$$

Formulas (2.47) imply that the expectation value of a product of two operators, where one is an arbitrary operator function $\mathbf{f}$, can be expressed solely in terms of the expectation value, f, and its derivatives.

2.3 Some Properties of Spin Coherent States

We introduce now spin CSs that were first considered in [36,37] and have found considerable utility in applications in nonlinear optics, quantum radiophysics and solid state physics [32-35]. These CSs will be used below when considering the time-evolution of quantum systems whose Hamiltonians include spin operators, in quasiclassical region of parameters. Consider the dimensionless spin operators $\mathbf{S}^z, \mathbf{S}^+, \mathbf{S}^-$ with the usual commutation relations,

$$[\mathbf{S}^+, \mathbf{S}^-] = 2\mathbf{S}^z, \qquad [\mathbf{S}^\pm, \mathbf{S}^z] = \mp \mathbf{S}^\pm, \tag{2.49}$$

$$[\mathbf{S}^\pm, \mathbf{S}^\pm] = 0, \qquad \mathbf{S}^\pm = \mathbf{S}^x \pm i\mathbf{S}^y.$$

For example, for spin 1/2, the operators in (2.49) could be expressed through the Pauli matrices,

$$\mathbf{S}^x = \frac{1}{2}\begin{vmatrix} 0 & 1 \\ 1 & 0 \end{vmatrix}, \qquad \mathbf{S}^y = \frac{1}{2}\begin{vmatrix} 0 & -i \\ i & 0 \end{vmatrix}, \qquad \mathbf{S}^z = \frac{1}{2}\begin{vmatrix} 1 & 0 \\ 0 & -1 \end{vmatrix}. \quad (2.50)$$

The spin CS $|\xi>$ [36,37] is a superposition of spin states $|S, M>$ (with $S^z = M; -S \leq M \leq S$) generated from the ground state $|S, -S>$ by

$$|\xi>= (1 + |\xi|^2)^{-S} \exp(\xi \mathbf{S}^+)|S, -S> . \qquad (2.51)$$

In this expression, ξ is an arbitrary complex number ($0 \leq |\xi| < \infty$) and the ground state $|S, -S>$ satisfies the following relations

$$\mathbf{S}^z|S, -S>= -S|S, -S>, \qquad (2.52)$$
$$\mathbf{S}^2|S, -S>= S(S+1)|S, -S>,$$

where

$$\mathbf{S}^2 = (\mathbf{S}^x)^2 + (\mathbf{S}^y)^2 + (\mathbf{S}^z)^2, \qquad (2.53)$$

and S is the dimensionless spin magnitude. We briefly list some properties of the spin CSs (2.51).

1). Expansion of the spin CS $|\xi>$ in a basis of states $|S, M>$ with a projection M ($M = -S, ..., S$) on the z-axis yieds [36,37]

$$|\xi>= \sum_{M=-S}^{S} U_M(\xi)|S, M>, \qquad (2.54)$$

where the expansion coefficients are given by

$$U_M(\xi) = (1 + |\xi|^2)^{-S} \left[\frac{2S}{(S+M)!(S-M)!} \right]^{1/2} \xi^{S+M} . \qquad (2.55)$$

2). Two spin CSs with different ξ are not orthogonal to each other, since

$$< \xi|\mu >= [(1 + |\xi|^2)(1 + |\mu|^2]^{-S}(1 + \xi^* \mu)^{2S} . \qquad (2.56)$$

3). Equations (2.54) and (2.55) imply the probability of finding the projection M in spin CS $|\xi>$,

$$P_M(\xi) \equiv |<M,S|\xi>|^2 = \frac{(2S)!|\xi|^{2(S+M)}}{(1+|\xi|^2)^{2S}(S+M)!(S-M)!}. \quad (2.57)$$

In spin CS $|\xi>$ the expectation value of the operator $\mathbf{S}^z$ is

$$S^z \equiv <\xi|\mathbf{S}^z|\xi> = -S\frac{1-|\xi|^2}{1+|\xi|^2}. \quad (2.58)$$

We introduce the following values

$$p = S - M, \quad \bar{p} = S - S^z, \quad (0 \le p, \bar{p} \le 2S), \quad (2.59)$$

and seek the distribution of the value p in spin CS $|\xi>$ for the condition

$$|\xi|^2 \approx \frac{2S}{\bar{p}} \gg 1. \quad (2.60)$$

Condition (2.60) means that the CS $|\xi>$ is close to the completely inverted state, $|S,S>$. For the condition (2.60) we have from (2.58) the Poisson distribution for the value p,

$$P_{\bar{p}}(p) = \frac{e^{-\bar{p}}\bar{p}^p}{p!}, \quad (\bar{p}/2S \ll 1). \quad (2.61)$$

4). Spin CSs minimize the Heisenberg uncertainty condition. Namely, in the relation (see (2.13) and (2.49))

$$<(\mathbf{S}^x)^2><(\mathbf{S}^y)^2> \ \ge \ \frac{1}{4}<\mathbf{S}^z>^2 \quad (2.62)$$

equality is reached for spin CSs.

5). The spin CS basis is a complete set [36,37], since

$$\int |\xi><\xi|d\mu_s(\xi) = \mathbf{I}, \quad (2.63)$$

where

$$d\mu_s(\xi) = \frac{(2S+1)}{\pi}\frac{d^2\xi}{(1+|\xi|^2)^2}.$$

6). In spin CSs there are several useful identities for expectation values that are analogous those for the boson CSs discussed above (see remark 8 for boson CSs). Suppose we have an arbitrary function **f** of spin operators,

$$\mathbf{f} = \mathbf{f}(\mathbf{S}^z, \mathbf{S}^+, \mathbf{S}^-). \tag{2.64}$$

Then, the following identities hold for expectation values (see Appendix A)

$$< \xi|\mathbf{S}^+\mathbf{f}|\xi >= (1+|\xi|^2)^{-2S}\left(2S\xi^* - \xi^{*2}\frac{\partial}{\partial \xi^*}\right)(1+|\xi|^2)^{2S} f, \tag{2.65}$$

$$< \xi|\mathbf{S}^-\mathbf{f}|\xi >= (1+|\xi|^2)^{-2S}\frac{\partial}{\partial \xi^*}(1+|\xi|^2)^{2S} f,$$

$$< \xi|\mathbf{S}^z\mathbf{f}|\xi >= (1+|\xi|^2)^{-2S}\left(\xi^*\frac{\partial}{\partial \xi^*} - S\right)(1+|\xi|^2)^{2S} f,$$

$$< \xi|\mathbf{f}\mathbf{S}^+|\xi >= (1+|\xi|^2)^{-2S}\frac{\partial}{\partial \xi}(1+|\xi|^2)^{2S} f,$$

$$< \xi|\mathbf{f}\mathbf{S}^-|\xi >= (1+|\xi|^2)^{-2S}\left(2S\xi - \xi^2\frac{\partial}{\partial \xi}\right)(1+|\xi|^2)^{2S} f,$$

$$< \xi|\mathbf{f}\mathbf{S}^z|\xi >= (1+|\xi|^2)^{-2S}\left(\xi\frac{\partial}{\partial \xi} - S\right)(1+|\xi|^2)^{2S} f.$$

As for the boson CSs, these identities will be useful in expressing quantum dynamics directly in terms of expectation values. Additional properties of boson and spin CSs will be presented as needed below.

3 The Exact C-Number Equation for Time-Dependent Expectation Values in Coherent States

There are several ways of considering the dynamical behavior of quantum expectation values. For example, one could derive the equation for the time-dependent wave function (or density matrix) then solve it in some set of basis states and take expectation values of the corresponding operators. This approach is at a disadvantage, though in its physical interpretation for quasiclassical asymptotics, since the classical limit for nonlinear Hamiltonians is singular (see below), and thus, is valid only for finite times. As we will discuss, this limit is singular (in the sence of perturbation theory of partial differential equations) even for the dynamics of expectation values. In dealing with such a singular limit we believe it is clearest from the physical viewpoint to go directly to the expectation-value dynamics, rather then try to interpret this singularity in the wave-function dynamics, which depends on the choice of basis. (Further discussion of this singular limit from the viewpoint of Wigner function appears below, in section 10). Going directly to the expectation value dynamics allows us to distinguish which part of this singularity is connected with measurable physical phenomena, and which is connected just with the choice of representation. This approach results in closed c-number equations for the time-dependent expectation values for rather arbitrary Hamiltonians that include boson and spin operators [20,40,41]. The initial conditions we consider are the boson and spin coherent states. However, for different initial conditions an additional procedure exists, involving averaging over the initial density matrix that largely overcomes this limitation. This averaging procedure is considered in section 5.2 (see also section 12.7).

In this section we present the "direct approach" to expectation-value dynamics in a general form and derive a closed set of c-number equations, both for expectation values and for quantum correlation functions for a rather general class of Hamiltonians. Later, these equations will be used to analyze specific quantum systems.

3.1 Exact C-Number Equation for Quantum Expectation Values: Time-Independent Boson Hamiltonians in Coherent States

For simplicity we consider a quantum system described by an arbitrary time-independent Hamiltonian $\mathbf{H}(\mathbf{a}^+, \mathbf{a})$ with only one degree of freedom. Later in this section we present the corresponding results for time-dependent Hamiltonians. Of course, systems with one degree of freedom are of no relevance in quantum chaos in which we will be interested later. However, the results we present here generalize easily to the case of many degrees of freedom (see section 3.6).

The Heisenberg equation for an arbitrary operator function $\mathbf{f} = \mathbf{f}\left(\mathbf{a}^+(t), \mathbf{a}(t)\right)$ is defined by

$$\dot{\mathbf{f}} = \frac{i}{\hbar}[\mathbf{H}, \mathbf{f}]. \tag{3.1}$$

The time-dependent expectation value of this operator at an arbitrary time t for the initial CS $|\alpha>$ (2.24) is given by

$$f(t) \equiv f(\alpha^*, \alpha; t) = < \alpha|\mathbf{f}\left(\mathbf{a}^+(t), \mathbf{a}(t)\right)|\alpha > . \tag{3.2}$$

We average the Heisenberg equation (3.1) over the initial CS $|\alpha>$, and obtain the dynamical equation

$$\dot{f} = \frac{i}{\hbar}\left(< \alpha|\mathbf{H}\mathbf{f}|\alpha > - < \alpha|\mathbf{f}\mathbf{H}|\alpha >\right). \tag{3.3}$$

We put the operators $\mathbf{H}$ and $\mathbf{f}$ into normal-ordering form in terms of the initial operators $\mathbf{a}^+$ and $\mathbf{a}$. Here and below "initial" means that an operator is given at the time $t = 0$ and denoted

$$\mathbf{a}^+ \equiv \mathbf{a}^+(t = 0), \qquad \mathbf{a} \equiv \mathbf{a}(t = 0). \tag{3.4}$$

The term "normal ordering" means that all operators $\mathbf{a}^+$ appear to the left, and all operators $\mathbf{a}$ appear to the right in any operator expression. For example,

$$\mathbf{H} = \sum_{m,n} H_{m,n}(\mathbf{a}^+)^m \mathbf{a}^n, \tag{3.5}$$

$$\mathbf{f}(t) = \sum_{k,l} f_{k,l}(t)(\mathbf{a}^+)^k \mathbf{a}^l, \tag{3.6}$$

where $H_{m,n}$ and $f_{k,l}(t)$ are expansion coefficients. Thus, the first term in the expectation value dynamics, equation (3.3), becomes

$$< \alpha|\mathbf{H}\mathbf{f}|\alpha > = \sum_{m,n}\sum_{k,l} H_{m,n} f_{k,l}(t) < \alpha|(\mathbf{a}^+)^m \mathbf{a}^n (\mathbf{a}^+)^k \mathbf{a}^l|\alpha > \tag{3.7}$$

$$= (\alpha^*)^m \alpha^l < \alpha|\mathbf{a}^n(\mathbf{a}^+)^k|\alpha > .$$

It is easy to prove (see Appendix A) that

$$< \alpha|\mathbf{a}^n(\mathbf{a}^+)^k|\alpha >= \exp(-|\alpha|^2)\frac{\partial^n}{\partial(\alpha^*)^n}(\alpha^*)^k \exp(|\alpha|^2) \tag{3.8}$$

$$= \exp(-|\alpha|^2)\frac{\partial^k}{\partial\alpha^k}\alpha^n \exp(|\alpha|^2).$$

Using (3.8) we derive from (3.7) the closed-form expression,

$$< \alpha|\mathbf{Hf}|\alpha >= \exp(-|\alpha|^2)\sum_{m,n}\sum_{k,l} H_{m,n} f_{k,l}(t)(\alpha^*)^m \alpha^l \frac{\partial^n}{\partial(\alpha^*)^n}(\alpha^*)^k \tag{3.9}$$

$$\times \exp(|\alpha|^2) = \exp(-|\alpha|^2) H\left(\alpha^*, \frac{\partial}{\partial\alpha^*}\right) f \exp(|\alpha^2|),$$

where $f \equiv f(t)$ is the c-number expectation value (3.2) we seek and the differential operator H has the functional form of the normal-ordered operator function $\mathbf{H}(\mathbf{a}^+, \mathbf{a})$, but with the substitution

$$\mathbf{a}^+ \rightarrow \alpha^*, \qquad \mathbf{a} \rightarrow \frac{\partial}{\partial\alpha^*}.$$

Analogously we have for the second term in (3.3)

$$< \alpha|\mathbf{fH}|\alpha >= \exp(-|\alpha|^2) H\left(\frac{\partial}{\partial\alpha}, \alpha\right) f \exp(|\alpha|^2). \tag{3.10}$$

Note that the differential operator H in (3.9) is a formal expression; yet it can still be manipulated consistantly by using the rule that it "does not differentiate its own variables". Actually the expressions (3.9),(3.10) are generalizations of the expressions given in (2.47). By combining (3.9) and (3.10) we derive from (3.3)

$$\dot{f} = \hat{K} f, \qquad f = (\alpha^*, \alpha; t), \tag{3.11}$$

with the initial value

$$f(0) = f(\alpha^*, \alpha),$$

where the partial differential operator $\hat{K}$ has the form

$$\hat{K} = \frac{i}{\hbar} \exp(-|\alpha|^2) \left[H\left(\alpha^*, \frac{\partial}{\partial \alpha^*}\right) - H\left(\frac{\partial}{\partial \alpha}, \alpha\right)\right] \exp(|\alpha|^2). \quad (3.12)$$

Equation (3.11) with definition (3.12) is the linear c-number partial differential equation we have sought for the time-dependent expectation value $f(t)$ in the initial CS $|\alpha>$.

3.2 Generalization for the Case of Any Number of Degrees of Freedom

Equation (3.11) generalizes immediately to the case of an arbitrary number of degrees of freedom. In this case we have

$$(\mathbf{a}^+, \mathbf{a}) \equiv (\mathbf{a}_1^+, \mathbf{a}_1, \mathbf{a}_2^+, \mathbf{a}_2, ...), \qquad [\mathbf{a}_n, \mathbf{a}_m^+] = \delta_{n,m}. \quad (3.13)$$

Then, the equation (3.11),(3.12) takes the form

$$\dot{f} = \hat{K} f, \qquad f = f(\boldsymbol{\alpha}^*, \boldsymbol{\alpha}; t) = f(\alpha_1^*, \alpha_2^*, ...; \alpha_1, \alpha_2, ...; t), \quad (3.14)$$

$$\hat{K} = \frac{i}{\hbar} \exp(-|\boldsymbol{\alpha}|^2) \left(H(\boldsymbol{\alpha}^*, \frac{\partial}{\partial \boldsymbol{\alpha}^*}) - H(\boldsymbol{\alpha}, \frac{\partial}{\partial \boldsymbol{\alpha}}) \right) \exp(|\boldsymbol{\alpha}|^2),$$

where the following definitions are introduced

$$\boldsymbol{\alpha} \equiv (\alpha_1, \alpha_2, ...), \qquad |\boldsymbol{\alpha}|^2 \equiv \sum_n |\alpha_n|^2. \quad (3.15)$$

In (3.14) the quantity f,

$$f(\boldsymbol{\alpha}^*, \boldsymbol{\alpha}; t) =< \boldsymbol{\alpha}| \mathbf{f} \left(\mathbf{a}_1^+(t), \mathbf{a}_1(t); \mathbf{a}_2^+(t), \mathbf{a}_2(t); ... \right) |\boldsymbol{\alpha} >, \quad (3.16)$$

is the expectation value of the operator function $\mathbf{f}$ in the CS defined in the product-state representation,

$$|\boldsymbol{\alpha} >= \prod_n |\alpha_n >, \qquad \mathbf{a}_n |\alpha_n >= \alpha_n |\alpha_n > . \quad (3.17).$$

3.3 Example: Dynamics of Expectation Values for a Nonlinear Quantum Oscillator

As an example we consider the quantum nonlinear oscillator with Hamiltonian

$$\mathbf{H}_0 = \hbar\omega\mathbf{a}^+\mathbf{a} + \hbar^2\mu(\mathbf{a}^+\mathbf{a})^2, \qquad (\mu > 0), \tag{3.18}$$

where ω is the frequency of linear oscillations and μ is the parameter of nonlinearity. For this case, (3.11) and (3.12) provide the following exact c-number equation for the time-dependent expectation value $f(t)$ (3.2)

$$\dot{f} = i\left(\omega + \mu\hbar + 2\mu\hbar|\alpha|^2\right)\left(\alpha^*\frac{\partial}{\partial\alpha^*} - \alpha\frac{\partial}{\partial\alpha}\right)f \tag{3.19}$$

$$+ i\mu\hbar\left((\alpha^*)^2\frac{\partial^2}{\partial(\alpha^*)^2} - \alpha^2\frac{\partial^2}{\partial\alpha^2}\right)f.$$

Equation (3.19) is expressible in terms of classical independent variables, action I and angle θ (instead of α and α^*), by using the canonical substitution

$$\alpha = \sqrt{I/\hbar}\exp(-i\theta), \tag{3.20}$$

where

$$\left|\frac{\partial(\sqrt{\hbar}\alpha, \sqrt{\hbar}\alpha^*)}{\partial(I,\theta)}\right| = 1.$$

Then, it is easily seen that the following relations hold,

$$\alpha^*\frac{\partial}{\partial\alpha^*} - \alpha\frac{\partial}{\partial\alpha} = -i\frac{\partial}{\partial\theta}, \tag{3.21}$$

$$\alpha^*\frac{\partial}{\partial\alpha^*} + \alpha\frac{\partial}{\partial\alpha} = 2I\frac{\partial}{\partial I},$$

$$(\alpha^*)^2\frac{\partial^2}{\partial(\alpha^*)^2} - \alpha^2\frac{\partial^2}{\partial\alpha^2} = i\frac{\partial}{\partial\theta} - 2iI\frac{\partial^2}{\partial I\partial\theta}.$$

Using these relations in (3.19) implies an equation for expectation value $f(I, \theta, t)$, namely,

$$\frac{\partial f}{\partial t} = (\omega + 2\mu I)\frac{\partial f}{\partial \theta} + 2\mu\hbar I \frac{\partial^2 f}{\partial I \partial \theta}. \tag{3.22}$$

Equation (3.22) involves the two dimensionless parameters that were mentioned in the Introduction:

$$\epsilon = \frac{\hbar}{I_0}, \quad \bar{\mu} = \frac{\mu I_0}{\omega}, \tag{3.23}$$

where $I_0 = \hbar|\alpha_0|^2$ is a characteristic action of the system. The quantum parameter ϵ becomes small ($\epsilon \ll 1$) in the quasiclassical region. (Later, we will call ϵ the "quasiclassical parameter".) The parameter $\bar{\mu}$ is the classical dimensionless parameter of nonlinearity, which is defined by

$$\bar{\mu} \equiv \frac{I}{\omega(I)} \frac{d\omega(I)}{dI} \sim \frac{H_{nonlinear}}{H_{linear}}, \tag{3.24}$$

where $\omega(I)$ is the frequency of classical nonlinear oscillations. In our case $\omega(I) = \omega + 2\mu I$. Next, we introduce dimensionless variables (y, τ) given by

$$\frac{I}{I_0} = e^y, \quad (0 \leq I < \infty, -\infty < y < \infty), \quad \tau = \omega t, \tag{3.25}$$

and write equation (3.22) in dimensionless form, as

$$\frac{\partial f}{\partial \tau} = (1 + 2\bar{\mu}e^y)\frac{\partial f}{\partial \theta} + 2\bar{\mu}\epsilon \frac{\partial^2 f}{\partial y \partial \theta}. \tag{3.26}$$

Setting $\epsilon = 0$ in (3.26) gives the classical equation for $f = f(y, \theta, \tau)$ which includes only first order derivatives. The quantum term involves the second order derivative, multiplied by the two parameters $\bar{\mu}$ and ϵ. So, the quantum term vanishes either in the linear case ($\bar{\mu} = 0$), or in the purely classical case ($\epsilon = 0$). However, in the general case ($\bar{\mu} \neq 0, \epsilon \neq 0$) the quantum term in (3.26) appears as a singular perturbation of the classical equation (because the small parameter ϵ multiplies the highest-order derivative). As we shall see, the presence of the quantum term causes the solution to drift away from its classical behavior as time proceeds. Thus, the classical limit exists as $\epsilon \to 0$ in the expectation-value dynamics, but the quantum and classical solutions drift apart over time because of dispersion due to the quantum term. Consequently - even when the initial population of the system with the Hamiltonian (3.18) lies in the deep quasiclassical region - the quantum-mechanical time-dependent expectation values will follow the corresponding classical solutions only for a finite time. This example shows that "quantum secularity" in the *expectation values* arises even in the simplest quantum nonlinear Hamiltonians. This secularity is of a physical nature and is not connected with any mathematical approximations or methods.

In the case at hand, the character of this secularity is easily seen by using separation of variables to solve equation (3.26). Write the function $f(y, \theta, \tau)$ in (3.26) in form

$$f(y, \theta, \tau) = \sum_{n=-\infty}^{\infty} f_n(y, \tau) e^{in\theta}. \tag{3.27}$$

Then, the function $f_n(y, \tau)$ satisfies

$$\frac{\partial f_n}{\partial \tau} = in(1 + 2\bar{\mu}e^y) f_n + 2i\bar{\mu}\epsilon n \frac{\partial f_n}{\partial y}, \tag{3.28}$$

which is easily integrated. For arbitrary initial function $f_n(y, 0)$ the solution of equation (3.28) is

$$f_n(y, \tau) = f_n(y + 2\bar{\mu}\epsilon i n\tau, 0) \exp\left[in\tau + \frac{e^y}{\epsilon}\left(e^{2in\bar{\mu}\epsilon\tau} - 1\right)\right]. \tag{3.29}$$

Substituting (3.29) into (3.27) gives the general solution of equation (3.26) in the form

$$f(y, \theta, \tau) = \sum_{n=-\infty}^{\infty} f_n(y + 2\bar{\mu}\epsilon i n\tau, 0) \exp\left[in\tau + \frac{e^y}{\epsilon}\left(e^{2in\bar{\mu}\epsilon\tau} - 1\right) + in\theta\right]. \tag{3.30}$$

The general solution (3.30) illustrates the character of the secularity injected by the quantum term in equation (3.26). The characteristic time-scale $\tau_\hbar$ at which the quasiclassical approach breaks down for integrable nonlinear boson and spin systems is discussed in section 5. We refer to section 5 for more analysis of this example, as well, only remarking now that: 1) the classical limit exists and is regular for $\epsilon = 0$; and 2) the quantum solution is quasiperiodic, not simple harmonic.

Also note that the "quantum secularity" for expectation values discussed here remains, if one makes an additional averaging over an initial density matrix. Consequently, this secularity is general and is not connected with the use of CSs as initial states (see section 5.2).

As was already mentioned equation (3.11) is linear, although it applies to a general class of nonlinear Hamiltonians. Next we re-express equation (3.11) in nonlinear form.

3.4 Nonlinear Representation of Expectation-Value Dynamics

We begin by considering a classical nonlinear Hamiltonian $H(p,q)$, where p and q are classical canonical variables. Hamiltonian's equations are

$$\dot{p} = \{H(p,q),p\} = -\frac{\partial H}{\partial q}, \qquad \dot{q} = \{H(p,q),q\} = \frac{\partial H}{\partial p}, \tag{3.31}$$

where $\{...\}$ are canonical Poisson brackets. Equations (3.31) are generally nonlinear. However, upon (formally) making a canonical transformation to initial values,

$$p_0 = p(t=0), \qquad q_0 = q(t=0),$$

Hamiltonian's equations for the functions $p(p_0, q_0, t)$ and $q(p_0, q_0, t)$ for the time-independent Hamiltonian $H(p,q) = H(p_0, q_0)$ take the linear form,

$$\dot{p} = \{H(p_0,q_0),p\} = \left[\frac{\partial H(p_0,q_0)}{\partial p_0}\frac{\partial}{\partial q_0} - \frac{\partial H(p_0,q_0)}{\partial q_0}\frac{\partial}{\partial p_0}\right] p, \tag{3.32}$$

$$\dot{q} = \{H(p_0,q_0),q\} = \left[\frac{\partial H(p_0,q_0)}{\partial p_0}\frac{\partial}{\partial q_0} - \frac{\partial H(p_0,q_0)}{\partial q_0}\frac{\partial}{\partial p_0}\right] q.$$

These equations are the classical analog of the linear quantum equation (3.11) and (3.12). Next, we shall consider the quantum analog of the nonlinear classical equations (3.31). First we note that the quantum c-number equation (3.11) and (3.12) may be written in the form (see Appendix A)

$$\dot{f} = \frac{i}{\hbar}\left\{ H \exp\left(\frac{\overleftarrow{\partial}}{\partial\alpha}\frac{\overrightarrow{\partial}}{\partial\alpha^*}\right) f - f \exp\left(\frac{\overleftarrow{\partial}}{\partial\alpha}\frac{\overrightarrow{\partial}}{\partial\alpha^*}\right) H \right\}. \tag{3.33}$$

In (3.33) the arrows indicate the direction of the action of the corresponding differential operators, and H and $f(t)$ are the c-number equivalents of the normally ordered operators (3.5) and (3.6) in CS $|\alpha>$

$$H =< \alpha|\mathbf{H}|\alpha >= \sum_{m,n} H_{m,n}(\alpha^*)^m \alpha^n, \tag{3.34}$$

$$f(t) =< \alpha|\mathbf{f}(t)|\alpha >= \sum_{k,l} f_{k,l}(t)(\alpha^*)^k \alpha^l.$$

Equation (3.33) is the c-number equivalent of the corresponding Heisenberg operator equation for an arbitrary operator $\mathbf{f}(t)$. Equation (3.33) follows from a general formula for the c-number equivalent for two arbitrary normally ordered operators $\mathbf{A}$ and $\mathbf{B}$ in CS $|\alpha>$ [31] (The proof is the same as we used in deriving equation (3.33) in Appendix A), namely

$$< \alpha|\mathbf{AB}|\alpha >= A \exp\left(\frac{\overleftarrow{\partial}}{\partial\alpha}\frac{\overrightarrow{\partial}}{\partial\alpha^*}\right) B. \tag{3.35}$$

Now we show how equation (3.33) leads to a nonlinear quantum c-number equation for $f(t)$ that has a classical limit of the type (3.31). We return to the example of the nonlinear oscillator with Hamiltonian (3.18) and choose $\alpha(t)$ for the function $f(t)$ in (3.11) and (3.33),

$$f(t) \equiv \alpha(t) = \alpha(\alpha_0^*, \alpha_0, t), \qquad \alpha(t=0) = \alpha_0. \tag{3.36}$$

Then, the Heisenberg equation for the operator $\mathbf{a}$ gives

$$i\hbar\dot{\alpha} =< \alpha_0|[\mathbf{a}, \mathbf{H}]|\alpha_0 >= \hbar(\omega + \mu\hbar)\alpha + 2\hbar^2\mu < \alpha_0|\mathbf{a}^+\mathbf{aa}|\alpha_0 > . \tag{3.37}$$

Let all operators $\mathbf{a}^+$ and $\mathbf{a}$ be normally ordered in the initial operators $\mathbf{a}_0^+$ and $\mathbf{a}_0$. Then, formula (3.35) yields

$$< \alpha_0|\mathbf{a}^+\mathbf{a}\mathbf{a}|\alpha_0 >= \alpha^* \exp\left(\frac{\overleftarrow{\partial}}{\partial\alpha_0}\frac{\overrightarrow{\partial}}{\partial\alpha_0^*}\right) \alpha \,\overleftarrow{\exp}\left(\frac{\overleftarrow{\partial}}{\partial\alpha_0}\frac{\overrightarrow{\partial}}{\partial\alpha_0^*}\right) \alpha. \quad (3.38)$$

Here, the left arrow in the symbol $\overleftarrow{\exp}$ means that this differential operator acts only on the nearest function on the left. Using (3.38) gives an exact nonlinear c-number equation for the quantum function $\alpha(t) \equiv \alpha(\alpha_0^*, \alpha_0, t)$,

$$i\dot{\alpha} = (\omega + \mu\hbar)\alpha + 2\hbar\mu\alpha^* \exp\left(\frac{\overleftarrow{\partial}}{\partial\alpha_0}\frac{\overrightarrow{\partial}}{\partial\alpha_0^*}\right) \alpha \,\overleftarrow{\exp}\left(\frac{\overleftarrow{\partial}}{\partial\alpha_0}\frac{\overrightarrow{\partial}}{\partial\alpha_0^*}\right) \alpha, \quad (3.39)$$

with initial condition $\alpha(t=0) = \alpha_0$. In the classical limit ($|\alpha_0| \to \infty, \hbar \to 0, |\alpha_0|^2\hbar \to I_0, finite$), equation (3.39) easily leads to

$$i\dot{\alpha} = \omega\alpha + 2\mu|\alpha|^2\alpha + O(\hbar^{5/2}), \qquad \alpha(t=0) = \alpha_0, \quad (3.40)$$

where, in taking this limit, we have multiplied equation (3.39) by $\sqrt{\hbar}$ and replaced $\sqrt{\hbar}\alpha \to \alpha$ (see (3.17). The expectation-value equations (3.11) and (3.33) (or (3.39)) will be useful in section 4 to formulate quasiclassical dynamical perturbation theory for expectation values.

3.5 Exact C-Number Equation for Quantum Expectation Values: Time-Independent Spin Hamiltonians in Coherent States

In analogy to section 3.1 for boson Hamiltonians, this section treats a quantum system with an arbitrary time-independent spin Hamiltonian $\mathbf{H}(\mathbf{S}^z, \mathbf{S}^+, \mathbf{S}^-)$ with one degree of freedom. Generalizing to the case of any number of spin degrees of freedom follows in analogy to section 3.2 for boson Hamiltonians. The Heisenberg equation for an arbitrary spin operator function $\mathbf{f}$ is

$$\dot{\mathbf{f}} = \frac{i}{\hbar}[\mathbf{H}, \mathbf{f}], \quad \mathbf{f} = \mathbf{f}\left(\mathbf{S}^z(t), \mathbf{S}^+(t), \mathbf{S}^-(t)\right). \quad (3.41)$$

The time-dependent expectation value $f(t)$ of the Heisenberg operator function $\mathbf{f}(t)$ is given by,

$$f(t) \equiv f(\xi^*, \xi, t) = < \xi|\mathbf{f}\left(\mathbf{S}^z(t), \mathbf{S}^+(t), \mathbf{S}^-(t)\right)|\xi > . \tag{3.42}$$

Averaging the Heisenberg equation (3.41) in the CS $|\xi >$ (2.51) gives

$$\dot{f} = \frac{i}{\hbar}(< \xi|\mathbf{Hf}|\xi > - < \xi|\mathbf{fH}|\xi >). \tag{3.43}$$

Next, we introduce a polynomial spin operator $\mathbf{H}$ in the form

$$\mathbf{H} = \sum_{m,l,n} A_{m,l,n}(\mathbf{S}^z)^m(\mathbf{S}^+)^l(\mathbf{S}^-)^n, \tag{3.44}$$

and consider the expression

$$< \xi|\mathbf{Hf}|\xi > = \sum_{m,l,n} A_{m,l,n} < \xi|(\mathbf{S}^z)^m(\mathbf{S}^+)^l(\mathbf{S}^-)^n\mathbf{f}|\xi > . \tag{3.45}$$

Using the identities (2.65) allows the expression $< \xi|...|\xi >$ in (3.45) to be expanded in the following way

$$< \xi|(\mathbf{S}^z)^m(\mathbf{S}^+)^l(\mathbf{S}^-)^n\mathbf{f}|\xi > \tag{3.46}$$

$$= (1+|\xi|^2)^{-2S}(\xi^*\frac{\partial}{\partial\xi^*} - S)^m(1+|\xi|^2)^{2S} < \xi|(\mathbf{S}^+)^l(\mathbf{S}^-)^n\mathbf{f}|\xi >,$$

where the effect of the spin operator $(\mathbf{S}^z)^m$ is expressed now as a differential operator. Analogously we treat the operators $(\mathbf{S}^+)^l$ and $(\mathbf{S}^-)^n$, thereby obtaining

$$< \xi|(\mathbf{S}^z)^m(\mathbf{S}^+)^l(\mathbf{S}^-)^n\mathbf{f}|\xi > = \tag{3.47}$$

$$(1+|\xi|^2)^{-2S}(\xi^*\frac{\partial}{\partial\xi^*}-S)^m(2S\xi^*-\xi^{*2}\frac{\partial}{\partial\xi^*})^l(\frac{\partial}{\partial\xi^*})^n(1+|\xi|^2)^{2S} < \xi|\mathbf{f}|\xi > .$$

Thus, the expression (3.45) can be represented in differential operator form as

$$< \xi|\mathbf{Hf}|\xi > \tag{3.48}$$

$$= (1+|\xi|^2)^{-2S} H\left[(\xi^*\frac{\partial}{\partial\xi^*} - S), (2S\xi^* - \xi^{*2}\frac{\partial}{\partial\xi^*}), (\frac{\partial}{\partial\xi^*})\right]$$

$$\times(1+|\xi|^2)^{2S} < \xi|\mathbf{f}|\xi > .$$

Likewise, the second term on the right side of (3.43) may be expressed as

$$< \xi|\mathbf{fH}|\xi > = \sum_{m,l,n} A_{m,l,n} < \xi|\mathbf{f}(\mathbf{S}^z)^m(\mathbf{S}^+)^l(\mathbf{S}^-)^n|\xi > \qquad (3.49)$$

$$\times(1+|\xi|^2)^{-2S} H\left[(\xi\frac{\partial}{\partial\xi}-S),(\frac{\partial}{\partial\xi}),(2S\xi-\xi^2\frac{\partial}{\partial\xi})\right]$$

$$(1+|\xi|^2)^{2S} < \xi|\mathbf{f}|\xi > .$$

As a result we derive from (3.43), (3.48), (3.38), (3.49) an exact quantum c-number equation for a time-dependent expectation value $f(t) \equiv < \xi|\mathbf{f}(t)|\xi >$

$$\dot{f} = \hat{K}f, \qquad f \equiv f(\xi^*,\xi,t), \qquad f(0) = f(\xi^*,\xi), \qquad (3.50)$$

where

$$\hat{K} = \frac{i}{\hbar}(1+|\xi|^2)^{-2S}\{H\left[(\xi^*\frac{\partial}{\partial\xi^*}-S),(2S\xi^*-\xi^{*2}\frac{\partial}{\partial\xi^*}),(\frac{\partial}{\partial\xi^*})\right]$$

$$-H\left[(\xi\frac{\partial}{\partial\xi}-S),(\frac{\partial}{\partial\xi}),(2S\xi-\xi^2\frac{\partial}{\partial\xi})\right]\}(1+|\xi|^2)^{2S}.$$

Equation (3.50) was first derived in [41]. In this equation the derivatives with respect to the CS parameters ξ and ξ^* act on *all* functions on the right, including the CS parameters in f. The generalization of this equation to the case of an arbitrary number of degrees of freedom follows easily.

3.6 Exact C-Number Equation for Quantum Time-Dependent Expectation Values for Boson-Spin Hamiltonians with Arbitrary Number of Degrees of Freedom

Very often problems arise with Hamiltonians that include boson and spin operators simultaneously. Moreover, such systems may possess many degrees of freedom (even an infinite numbers). In such cases we must generalize c-number equations (3.11),(3.12) and (3.50) derived above. Consider a time-independent Hamiltonian

$$\mathbf{H} = \mathbf{H}\left(\{\mathbf{a}^+\}, \{\mathbf{a}\}, \{\mathbf{S}^z\}, \{\mathbf{S}^+\}, \{\mathbf{S}^-\}\right) \tag{3.51}$$

$$= \sum A^{j_1,j_2,j_3,j_4,j_5}_{i_1,i_2,i_3,i_4,i_5} (\mathbf{a}^+_{i_1})^{j_1} (\mathbf{a}_{i_2})^{j_2} (\mathbf{S}^z_{i_3})^{j_3} (\mathbf{S}^+_{i_4})^{j_4} (\mathbf{S}^-_{i_5})^{j_5},$$

where we have introduced the notation $\{\mathbf{A}\} \equiv \mathbf{A}_1, \mathbf{A}_2, \ldots$. Define at $t = 0$ a CS

$$|\{\alpha\}, \{\xi\} >= \prod_n \prod_m |\alpha_n > |\xi_m >, \tag{3.52}$$

where $|\alpha_n >$ and $|\xi_m >$ are the CSs introduced before, (2.24) and (2.51), respectively. We also introduce the time-dependent expectation value of an arbitrary operator

$$\mathbf{f}(t) \equiv \mathbf{f}\left(\{\mathbf{a}^+(t)\}, \{\mathbf{a}(t)\}, \{\mathbf{S}^z(t)\}, \{\mathbf{S}^-(t)\}, \{\mathbf{S}^-(t)\}\right) \tag{3.53}$$

in CS (3.52), as

$$f(t) \equiv < \{\xi\}, \{\alpha\} | \mathbf{f}(t) | \{\alpha\}, \{\xi\} > . \tag{3.54}$$

Then, we have the c-number equation for $f(t)$ (3.54)

$$\dot{f} = \hat{K} f, \qquad f = f(\{\alpha^*\}, \{\alpha\}, \{\xi^*\}, \{\xi\}, t), \tag{3.55}$$

where the operator $\hat{K}$ has the form

$$\hat{K} = \frac{i}{\hbar} \exp(-\sum_n |\alpha_n|^2) \prod_m (1 + |\xi_m|^2)^{-2S} \tag{3.56}$$

$$\{H\left(\{\alpha^*, \frac{\partial}{\partial \alpha^*}, \xi^* \frac{\partial}{\partial \xi^*} - S, 2S\xi^* - \xi^{*2} \frac{\partial}{\partial \xi^*}, \frac{\partial}{\partial \xi^*}\}\right)$$

$$-H\left(\{\alpha, \frac{\partial}{\partial \alpha}, \xi \frac{\partial}{\partial \xi} - S, \frac{\partial}{\partial \xi}, 2S\xi - \xi^2 \frac{\partial}{\partial \xi}\}\right)\}$$

$$\prod_{m'} (1 + |\xi_{m'}|^2)^{2S} \exp(\sum_{n'} |\alpha_{n'}|^2).$$

Equation (3.55) with definition (3.56) will be used in the following. In the next part of this section we shall derive a c-number equation for time-dependent quantum correlation functions.

3.7 Dynamics of Quantum Correlation Functions

Consider two arbitrary operators **F** and **G** and define their quantum time-dependent correlation function (QCF) in CS as

$$P_{F,G}(t) = <\xi,\alpha|\mathbf{F}(t)\mathbf{G}(t)|\alpha,\xi> \tag{3.57}$$
$$- <\xi,\alpha|\mathbf{F}(t)|\alpha,\xi><\xi,\alpha|\mathbf{G}(t)|\alpha,\xi> .$$

The QCF (3.57) vanishes in the classical limit since $P^{(cl)}_{F,G} = 0$ at any $t \neq 0$. Note that all three terms in the right side in (3.57) satisfy the equations

$$\frac{\partial <FG>}{\partial t} = \hat{K} <FG>, \qquad \frac{\partial F}{\partial t} = \hat{K}F, \qquad \frac{\partial G}{\partial t} = \hat{K}G, \tag{3.58}$$

where

$$<FG> \equiv <\xi,\alpha|\mathbf{F}(t)\mathbf{G}(t)|\alpha,\xi>, \quad F \equiv <\xi,\alpha|\mathbf{F}(t)|\alpha,\xi>, \tag{3.59}$$
$$G \equiv <\xi,\alpha|\mathbf{G}(t)|\alpha,\xi> .$$

Therefore, equations (3.58) and (3.59) imply a c-number equation for QCF $P_{F,G}(t)$ in (3.57), namely

$$\frac{\partial P_{F,G}(t)}{\partial t} = \hat{K}P_{F,G}(t) + \hat{K}FG - F\hat{K}G - G\hat{K}F. \tag{3.60}$$

In this equation the c-number functions $F = F(t)$ and $G = G(t)$ are supposed to be the solutions of the corresponding equations in (3.58). The QCF dynamical equation (3.60) can be simplified in different cases and will be illustrated in concrete examples in sections 8, 11-14.

3.8 Generalization of C-Number Equations in the Case of Time-Dependent Hamiltonians

In the case of a time-dependent Hamiltonian $\mathbf{H}(t)$ the direct method of deriving closed c-number equations for the time-dependent expectation value $f(t)$ cannot be applied. Nevertheless, we present here a variant of this method that can be applied in the case of a time-dependent Hamiltonian. Consider, for example, a Hamiltonian which depends on time periodically

$$\mathbf{H} = \mathbf{H}_0 + \lambda e^{i\omega t}\mathbf{V} + \lambda e^{-i\omega t}\mathbf{V}^+, \tag{3.61}$$

where $\mathbf{H}_0$ is time-independent Hamiltonian and λ is an interaction parameter. Next, introduce a new time-independent Hamiltonian,

$$\mathbf{H}_{eff} = \hbar\omega\mathbf{b}^+\mathbf{b} + \mathbf{H}_0 + \varepsilon(\mathbf{b}^+\mathbf{V} + \mathbf{b}\mathbf{V}^+). \tag{3.62}$$

In (3.62) ε is a parameter that will be discussed below, and $\mathbf{b}^+$, $\mathbf{b}$ are new boson operators ($[\mathbf{b}, \mathbf{b}^+] = 1$). For the operator $\mathbf{b} = \mathbf{b}(t=0)$ introduce a CS $|\beta>$, such that

$$\mathbf{b}|\beta>=\beta|\beta>. \tag{3.63}$$

Define the average number of quanta in the field "β" at $t = 0$ in the CS $|\beta>$ by

$$n_\beta^{(0)} =<\beta|\mathbf{b}^+\mathbf{b}|\beta>=|\beta|^2. \tag{3.64}$$

Assume that $|\beta| \gg 1$, so the field described by the operators $\mathbf{b}^+, \mathbf{b}$ is "strong". Then, one can see that in the limit

$$\beta = \beta_0 \to \infty, \quad \varepsilon \to 0, \quad \beta_0\varepsilon = \lambda = constant, \tag{3.65}$$

the Hamiltonian $\mathbf{H}_{eff}$ (3.62) transforms into the Hamiltonian (3.61) (for simplicity β_0 is taken to be real). Since the Hamiltonian $\mathbf{H}_{eff}$ does not depend explicitly on time, we can apply to $\mathbf{H}_{eff}$ the method discussed above and then make the transition (3.65). This method of treating periodically time-dependent Hamiltonians will be used below in sections 8 and 12.

4 Quasiclassical Perturbation Theory for Time-Dependent Expectation Values with Quantum Nonlinear Hamiltonians

We are dealing with quasiclassical dynamics of expectation values for nonlinear Hamiltonians. In general, the classical limit for time-dependent expectation values is valid only for the finite time interval $0 < t < \tau_\hbar$. This is because the quantum terms are singular perturbations of the classical dynamics. Indeed, in the corresponding c-number equations for expectation values the "small quantum parameter" $\hbar$ multiplies the highest order derivatives (see, for example, the equation (3.26)). So, in general, any perturbation theory gotten by an expansion in the Planck constant $\hbar$ (or, equivalently, the quasiclassical parameter ϵ) ,will be secular, and thus will be valid only for finite time. The results of this kind of perturbation theory will be verified in what follows for cases of nonlinear integrable boson and spin Hamiltonians like (3.18) (see section 5). We also varify this perturbation theory for the nonintegrable systems we treat by calculating their additional integrals of motion in both the classical and quantum cases. In this section we consider this kind of secular perturbation theory for the linear c-number equations of the type (3.11) and (3.50), and for nonlinear equations of the type (3.39). Further we apply the results of this section to derive the time-scale $\tau_\hbar$ and the dynamics of quantum correlation functions for different systems.

4.1 Quasiclassical Dynamical Perturbation Theory for Linear C-Number Equations

As a starting point we shall consider here the linear equations (3.11) and (3.50) for the expectation value $f(t)$ of boson and spin systems and decompose the differential operator $\hat{K}$ in (3.12) (or (3.50)) in the form

$$\hat{K} = \hat{K}_{cl} + \hbar \hat{K}_q, \tag{4.1}$$

where $\hat{K}_{cl}$ includes only first order derivatives (see concrete examples in sections 5, 11-14) and describes the corresponding classical limit. The other operator $\hat{K}_q$ in (4.1) includes higher order derivatives and is responsible for the quantum effects. To find the dynamics of quantum corrections at first order in $\hbar$, we express the solution of the quantum expectation value in the form

$$f(t) = f_{cl}(t) + \hbar f_q(t), \tag{4.2}$$

where $f_{cl}(t)$ is the solution of the corresponding classical equation

$$\frac{\partial f_{cl}(t)}{\partial t} = \hat{K}_{cl}(t) f_{cl}(t), \qquad f_{cl}(0) = f_{cl}(\alpha^*, \alpha; \xi^*, \xi). \tag{4.3}$$

Then, quantum correction $f_q(t)$ satisfies the exact equation

$$\frac{\partial f_q(t)}{\partial t} = \hat{K}_{cl} f_q(t) + \hat{K}_q f_{cl}(t) + \hbar \hat{K}_q f_q(t), \qquad f_q(0) = f_q(\alpha^*, \alpha; \xi^*, \xi). \tag{4.4}$$

In (4.4) the classical function $f_{cl}(t)$ is supposed to be known from the solution of the classical equation (4.3). At first order in $\hbar$, we shall neglect the last term in (4.4). Then, we have the following approximate equation for the quantum correction $f_q(t)$

$$\frac{\partial f_q(t)}{\partial t} = \hat{K}_{cl} f_q(t) + \hat{K}_q f_{cl}(t), \qquad f_q(0) = f_q(\alpha^*, \alpha; \xi^*, \xi). \tag{4.5}$$

Since the last term in (4.5) is known, equation (4.5) is a partial differential equation of the first order and can be solved, say, by using the method of characteristics [42]. Keeping in mind the notations for quantum systems with many degrees of freedom we present equation (4.5) in the standard form

$$\frac{\partial U(\vec{x}, t)}{\partial t} = \hat{L} U(\vec{x}, t) + D(\vec{x}, t), \qquad U(\vec{x}, 0) = U(\vec{x}), \tag{4.6}$$

where $\hat{L}$ is a differential operator that includes only first order derivatives,

$$\hat{L} = \sum_{j=1}^{N} L_j(\vec{x}) \frac{\partial}{\partial x_j}. \tag{4.7}$$

N is the number of degrees of freedom and $D(\vec{x}, t)$ is supposed to be a known function. We present the solution of equation (4.6) in explicit form, by introducing characteristic coordinates y_j given by

$$y_j = \varphi_j(\vec{x}, t), \qquad (j = 1, ..., N). \tag{4.8}$$

At $t = 0$ the new coordinates coincide with the old ones

$$\varphi_j(\vec{x}, 0) = x_j, \qquad (j = 1, ..., N), \tag{4.9}$$

and the functions $\varphi_j(\vec{x}, t)$ satisfy the equations

$$\frac{d\varphi_j(\vec{x}, t)}{dt} = L_j\left[\vec{\varphi}(\vec{x}, t)\right], \qquad \varphi_j(\vec{x}, 0) = x_j, \qquad (j = 1, ..., N). \tag{4.10}$$

In the equations (4.10) x_j is the initial coordinate and $y_j = \varphi_j(\vec{x}, t)$ is the current one. Express now the solution $U(\vec{x}, t)$ of (4.6) using the substitution (4.8). We have

$$U(\vec{x}, t) = U\left[\vec{\varphi}^{-1}(\vec{y}, t), t\right] \equiv \tilde{U}(\vec{y}, t), \tag{4.11}$$

and

$$\frac{\partial \tilde{U}(\vec{y}, t)}{\partial t} = \frac{\partial U(\vec{x}, t)}{\partial t} + \sum_j \frac{\partial U(\vec{x}, t)}{\partial x_j} \frac{\partial \varphi_j^{-1}(\vec{y}, t)}{\partial t}. \tag{4.12}$$

In (4.11), (4.12)

$$x_j = \varphi_j^{-1}(\vec{y}, t), \qquad \left(y_j = \varphi_j^{-1}(\vec{y}, 0)\right), \qquad (j = 1, ..., N), \tag{4.13}$$

and φ_j^{-1} are the inverse functions which determine the dependence of the initial coordinates through the final ones and the time. These inverse functions always exist and are single-valued for reversible Hamiltonian systems. For these functions we have

$$\frac{d\vec{\varphi}^{-1}(\vec{y}, t)}{dt} = -\hat{L}\left[\vec{\varphi}^{-1}(\vec{y}, t)\right], \qquad \left(\vec{\varphi}^{-1}(\vec{y}, 0) = \vec{y}\right). \tag{4.14}$$

Substituting (4.12) into (4.6) and taking into account (4.13) we find

$$\frac{\partial \tilde{U}(\vec{y},t)}{\partial t} - \sum_j \left\{ \frac{\partial U(\vec{x},t)}{\partial x_j} \frac{\partial \varphi_j^{-1}(\vec{y},t)}{\partial t} + L_j \left[\vec{\varphi}^{-1}(\vec{y},t)\right] \frac{\partial U(\vec{x},t)}{\partial x_j} \right\}$$
$$= D\left[\vec{\varphi}^{-1}(\vec{y},t),t\right]. \tag{4.15}$$

Since according to (4.14) the expression in the brackets {...} in (4.15) is equal to zero, we have from (4.15)

$$\frac{\partial \tilde{U}(\vec{y},t)}{\partial t} = D\left[\vec{\varphi}^{-1}(\vec{y},t),t\right]. \tag{4.16}$$

Integrating (4.16) we derive the solution of (4.6) in general form,

$$\tilde{U}(\vec{y},t) = U(\vec{x},t) = \int_0^t D\left[\vec{\varphi}^{-1}(\vec{y},\tau),\tau\right] d\tau. \tag{4.17}$$

The equation (4.17) gives an explicit form for the first order quantum correction by $\hbar$ to the classical solution. Further we shall use formula (4.17) for the different concrete systems.

4.2 Nonlinear Dynamical Perturbation Theory for Time-Dependent Expectation Values

We present here a nonlinear variant of quasiclassical dynamical perturbation theory. For simplicity consider a Hamiltonian which includes only boson operators

$$\mathbf{H} = \sum_{m,n\geq 0} \left[H_{m,n}(t)(\mathbf{a}^+)^m \mathbf{a}^n + h.c.\right], \qquad [\mathbf{a},\mathbf{a}^+] = \hbar, \tag{4.18}$$

where $H_{m,n}(t)$ are the coefficients of expansion which now can be time-dependent (in (3.5) the corresponding coefficients were time-independent), and for convenience the operators $\mathbf{a}^+$ and $\mathbf{a}$ are normalized by $\sqrt{\hbar}$. Define at $t=0$ a CS $|\alpha_0>$

$$\mathbf{a}_0|\alpha_0> = \alpha_0|\alpha_0>, \qquad (\mathbf{a}_0 \equiv \mathbf{a}(t=0)). \tag{4.19}$$

Consider the solution of the Heisenberg equation for the operator $\mathbf{a}(t)$

$$\mathbf{a}(t) = \mathbf{a}(\mathbf{a}_0^+, \mathbf{a}_0, t) = \sum_{m,n} \gamma_{m,n}(t)(\mathbf{a}_0^+)^m(\mathbf{a}_0)^n, \tag{4.20}$$

where $\gamma_{m,n}(t)$ are the coefficients of expansion. Then, we have for the expectation values $\alpha(t)$ and $\alpha^*(t)$ in the CS $|\alpha_0>$

$$\alpha(t) \equiv \alpha(\alpha_0^*, \alpha_0, t) \equiv <\alpha_0|\mathbf{a}(t)|\alpha_0> = \mathbf{a}^{(N)}(\alpha_0^*, \alpha_0, t), \tag{4.21}$$

$$\alpha^*(t) \equiv \alpha^*(\alpha_0^*, \alpha_0, t) \equiv <\alpha_0|\mathbf{a}^+(t)|\alpha_0> = \mathbf{a}^{+(N)}(\alpha_0^*, \alpha_0, t),$$

where the symbol (N) denotes that any operator function $\hat{Q}$ is normally ordered by the operators $\mathbf{a}_0^+$ and $\mathbf{a}_0$

$$\mathbf{Q}(\mathbf{a}_0^+, \mathbf{a}_0, t) = \sum_{m,n} Q_{m,n}(t)(\mathbf{a}_0^+)^m(\mathbf{a}_0)^n \equiv \mathbf{Q}^{(N)}(\mathbf{a}_0^+, \mathbf{a}_0, t). \tag{4.22}$$

To simplify the expressions the dependence on time in the operators $\mathbf{a}^+(t)$ and $\mathbf{a}(t)$ will not be denoted below. Using the definitions (4.21) we have from the the Heisenberg equation for the operator $\mathbf{a}(t)$

$$i\hbar\dot{\alpha} = \sum_{m,n}[mH_{m,n}(t) < \alpha_0|(\mathbf{a}^+)^{m-1}\mathbf{a}^n|\alpha_0> \tag{4.23}$$

$$+nH_{m,n}^*(t) < \alpha_0|(\mathbf{a}^+)^{n-1}\mathbf{a}^m|\alpha_0> .$$

In (4.23) all operators $\mathbf{a}^+$ and $\mathbf{a}$ are the functions on time and can be considered as normally ordered operator functions by the initial operators $\mathbf{a}_0^+$ and $\mathbf{a}_0$ (see (4.20)). Thus, the problem arises on calculating (or normal ordering) a product

$$\mathbf{Q}(\mathbf{a}_0^+, \mathbf{a}_0) = (\mathbf{a}^+)^p\mathbf{a}^q \tag{4.24}$$

of normally ordered operators $\mathbf{a}^+$ and $\mathbf{a}$ by the initial operators $\mathbf{a}_0^+$ and $\mathbf{a}_0$. Actually this problem was solved already by the formula (3.35). However, here we need only a quasiclassical approximation of this formula in the first order by $\hbar$. For this particular case we present here a direct "quasiclassical solution" of this problem. Introduce the normal ordering operator $\hat{N}$ [33]

$$\hat{N}\{(\mathbf{a}_0)^n(\mathbf{a}_0^+)^m\} = (\mathbf{a}_0^+)^m\mathbf{a}_0^n, \tag{4.25}$$

and use the formula

$$(\mathbf{a}_0)^n(\mathbf{a}_0^+)^m = \hat{N}\{(\mathbf{a}_0 + \hbar\frac{\partial}{\partial \mathbf{a}_0^+})^n(\mathbf{a}_0^+)^m\} \tag{4.26}$$

$$= (\mathbf{a}_0^+)^m\mathbf{a}_0^n + \hbar mn(\mathbf{a}_0^+)^{m-1}\mathbf{a}_0^{n-1} + O(\hbar^2),$$

which follows from (3.35). Let the operators $\mathbf{Q}_1$ and $\mathbf{Q}_2$ be the operator functions dependent on the operators $\mathbf{a}_0^+$ and $\mathbf{a}_0$

$$\mathbf{Q}_1 = \mathbf{Q}_1(\mathbf{a}_0^+, \mathbf{a}_0, t) = \mathbf{Q}_1^{(N)}(\mathbf{a}_0^+, \mathbf{a}_0, t) = \sum_{m,n} Q_{m,n}^{(1)}(t)(\mathbf{a}_0^+)^m(\mathbf{a}_0)^n,$$

$$\mathbf{Q}_2 = \mathbf{Q}_2(\mathbf{a}_0^+, \mathbf{a}_0, t) = \mathbf{Q}_2^{(N)}(\mathbf{a}_0^+, \mathbf{a}_0, t) = \sum_{m,n} Q_{m,n}^{(2)}(t)(\mathbf{a}_0^+)^m(\mathbf{a}_0)^n. \tag{4.27}$$

Present a quasiclassical expression for the normal ordering product $\mathbf{Q}_1\mathbf{Q}_2$. We have from (4.26), (4.27)

$$\mathbf{Q}_1\mathbf{Q}_2 = \mathbf{Q}_1^{(N)}\mathbf{Q}_2^{(N)} \tag{4.28}$$

$$= \sum_{m_1,n_1}\sum_{m_2,n_2} Q_{m_1,n_1}^{(1)}(t)Q_{m_2,n_2}^{(2)}(t)(\mathbf{a}_0^+)^{m_1}(\mathbf{a}_0)^{n_1}(\mathbf{a}_0^+)^{m_2}(\mathbf{a}_0)^{n_2}$$

$$= \hat{N}\{\mathbf{Q}_1^{(N)}\mathbf{Q}_2^{(N)} + \hbar\hat{D}(\mathbf{Q}_1^{(N)}, \mathbf{Q}_2^{(N)})\} + O(\hbar^2),$$

where the operator $\hat{D}$ is defined by the expression

$$\hat{D}(\mathbf{A}, \mathbf{B}) \equiv \frac{\partial \mathbf{A}}{\partial \mathbf{a}_0}\frac{\partial \mathbf{B}}{\partial \mathbf{a}_0^+}. \tag{4.29}$$

Using (4.20), (4.27), (4.28) we derive the quasiclassical formula for the normal ordered operator function $\mathbf{Q}$ (4.24)

$$\mathbf{Q}(\mathbf{a}_0^+, \mathbf{a}_0, t) = [\mathbf{a}^+(\mathbf{a}_0^+, \mathbf{a}_0, t)]^p[\mathbf{a}(\mathbf{a}_0^+, \mathbf{a}_0, t)]^q \tag{4.30}$$

$$= \hat{N}\{(\mathbf{a}^+)^p\mathbf{a}^q + \hbar pq\hat{D}(\mathbf{a}^+, \mathbf{a})(\mathbf{a}^+)^{p-1}\mathbf{a}^{q-1}$$

$$+\frac{\hbar}{2}p(p-1)\hat{D}(\mathbf{a}^+, \mathbf{a}^+)(\mathbf{a}^+)^{p-2}\mathbf{a}^q + \frac{\hbar}{2}q(q-1)\hat{D}(\mathbf{a}, \mathbf{a})(\mathbf{a}^+)^p\mathbf{a}^{q-2}\} + O(\hbar^2).$$

Taking into account the definition (4.25), we get from (4.30)

$$< \alpha_0|(\mathbf{a}^+)^p\mathbf{a}^q|\alpha_0 >= (\alpha^*)^p\alpha^q + \hbar pq\hat{D}(\alpha^*, \alpha)(\alpha^*)^{p-1}\alpha^{q-1} \tag{4.31}$$

$$+\frac{\hbar}{2}p(p-1)\hat{D}(\alpha^*,\alpha^*)(\alpha^*)^{p-2}\alpha^q+\frac{\hbar}{2}q(q-1)\hat{D}(\alpha,\alpha)(\alpha^*)^p\alpha^{q-2}\}+0(\hbar^2),$$

where analogously to (4.29)

$$\hat{D}(A,B)\equiv\frac{\partial A}{\partial\alpha_0}\frac{\partial B}{\partial\alpha_0^*}. \tag{4.32}$$

In (4.32) A and B are the c-number functions. Substituting (4.32) in (4.23) we derive the equation for $\alpha(\alpha_0^*,\alpha_0,t)$

$$\begin{aligned}
i\dot{\alpha} = &\sum_{m,n}[mH_{m,n}(t)(\alpha^*)^{n-1}\alpha^n + nH_{m,n}(t)(\alpha^*)^{n-1}\alpha^m] \qquad (4.33)\\
&+\hbar\sum_{m,n}\{H_{m,n}(t)[m(m-1)n(\alpha^*)^{m-2}\alpha^{n-1}\hat{D}(\alpha^*,\alpha)\\
&+\frac{1}{2}m(m-1)(m-2)(\alpha^*)^{m-3}\alpha^n\hat{D}(\alpha^*,\alpha^*)\\
&+\frac{1}{2}mn(n-1)(\alpha^*)^{m-1}\alpha^{n-2}\hat{D}(\alpha,\alpha)]\\
&+H^*_{m,n}(t)[mn(n-1)(\alpha^*)^{n-2}\alpha^{m-1}\hat{D}(\alpha^*,\alpha)\\
&+\frac{1}{2}n(n-1)(n-2)(\alpha^*)^{n-3}\alpha^m\hat{D}(\alpha^*,\alpha^*)\\
&+\frac{1}{2}mn(m-1)(\alpha^*)^{n-1}\alpha^{m-2}\hat{D}(\alpha,\alpha)]\}+O(\hbar^2).
\end{aligned}$$

The equation (4.33) is a closed nonlinear c-number ordinary differential equation of motion in "quantum phase space" (α,α^*) in the quasiclassical approximation. The terms in the first square brackets in (4.33) describe the classical effects, and the terms $\hbar\{...\}$ describe the quantum corrections. Consider as an example the Hamiltonian (3.18), which in the notations used in the section (4.2) has the form

$$\mathbf{H}=\omega\mathbf{a}^+\mathbf{a}+\mu(\mathbf{a}^+\mathbf{a})^2, \qquad (\mu>1), \qquad [\mathbf{a}^+,\mathbf{a}]=\hbar. \tag{4.34}$$

In this case we have for non zero coefficients $H_{m,n}$ in (4.18)

$$H_{1,1}=H^*_{1,1}=\frac{\omega+\hbar\mu}{2}, \qquad H_{2,2}=H^*_{2,2}=\frac{\mu}{2}.$$

Then, we have for this case from (4.33) the quasiclassical nonlinear c-number equation for $\alpha(t) \equiv \alpha(\alpha_0^*, \alpha_0, t)$

$$i\dot{\alpha} = [(\omega + \hbar\mu) + 2\mu|\alpha|^2]\alpha + 2\mu\hbar[\hat{D}(\alpha, \alpha)\alpha^* + 2\hat{D}(\alpha^*, \alpha)\alpha]. \quad (4.35)$$

The same equation follows from the exact nonlinear equation (3.39) in the quasiclassical approximation, as in this case we have for the last term in (3.39) (in the notations where $\alpha \sim \sqrt{\hbar}$)

$$\alpha^* \exp\left(\frac{\overleftarrow{\partial}}{\partial \alpha_0}\frac{\overrightarrow{\partial}}{\partial \alpha_0^*}\right) \alpha \overleftarrow{\exp}\left(\frac{\overleftarrow{\partial}}{\partial \alpha_0}\frac{\overrightarrow{\partial}}{\partial \alpha_0}\right) \alpha \quad (4.36)$$

$$= |\alpha|^2\alpha + \hbar\left(2\frac{\partial \alpha^*}{\partial \alpha_0}\frac{\partial \alpha}{\partial \alpha_0^*}\alpha + \frac{\partial \alpha}{\partial \alpha_0}\frac{\partial \alpha}{\partial \alpha_0^*}\alpha^*\right) + O(\hbar^2).$$

Note that an advantage of the quasiclassical approach considered in section 4.2 consists in its applicability for any time-dependent nonlinear Hamiltonian of the type (4.18). Nevertheless, the functional dependence (4.20) is rather complicated

$$\mathbf{a}(t) = \hat{U}^{-1}(t)\mathbf{a}_0\hat{U}(t), \quad (4.37)$$

where $\hat{U}(t)$ is the evolution operator of the wave function $|\Psi(t)>$

$$|\Psi(t)> = \hat{U}(t)|\Psi_0>, \qquad (|\Psi_0> = |\alpha_0>), \quad (4.38)$$

the c-number equation (4.33) allows to calculate directly the time-dependent expectation value $\alpha(t)$ in quasiclassical approximation without preliminary calculation of the evolution operator $\hat{U}(t)$ in (4.37), (4.38). We shall apply this approach when considering the problem of quantum chaos: the quantum system with an infinite number of interacting quantum nonlinear resonances-quantum nonlinear oscillator (3.18) interacting with a periodic in time sequence of δ-impulses (section 6.1), and the boson system of two interacting quantum nonlinear resonances (section 6.2).

5 The Characteristic Times of Violation of Quasiclassical Approximation for Integrable Boson and Spin Systems

In the previous sections we were mainly interested in considering the methods which can allow us to analyse the quantum corrections to the classical solutions for different boson and spin systems. As a first step we consider in this section the application of the considered above c-number equations for quantum integrable boson and spin systems in quasiclassical region of parameters. We derive one of the main characteristic dynamical parameter, namely the characteristic time-scale $\tau_\hbar$ of violation of the quasiclassical solutions. For this we introduce below in section 5.1 a function $\delta(t)$ which characterises the difference in time between quantum and corresponding classical solutions. It will be shown that in general case the universal time-scale $\tau_\hbar$ which depends only on the parameters of the Hamiltonian does not exist. Namely, the time-scale $\tau_\hbar$ depends also on the expectation value one measures, and, for example, for big momentums this time-scale $\tau_\hbar$ can significantly decrease. Nevertheless, for the typical cases of not big momentums there exists rather universal time-scale $\tau_\hbar$ which can be expressed in the terms of two mentioned above main parameters - classical parameter of nonlinearity $\bar{\mu}$ and a quasiclassical parameter $\kappa = I/\hbar = 1/\epsilon$ (where I is a characteristic action of the system): $\tau_\hbar \sim \sqrt{\kappa}/\bar{\mu}$. In section 5.2 we investigate the problem of time-scale $\tau_\hbar$ taking into account an additional averaging over the initial density matrix. It is shown that when the initial distribution is rather narrow the results for time-scale $\tau_\hbar$ are mainly the same as for CSs. In spite of this there are some peculiarities connected with the additional averaging which are also discussed in section 5.2. In section 5.3 the quantum correlation function is analyzed. In section 5.4 the time-scale $\tau_\hbar$ is considered for integrable nonlinear spin system in CSs. It is shown that the characteristic value for $\tau_\hbar$ in this case also can be expressed in the terms of two parameters $\bar{\mu}$ and κ and has the same form as for boson system. In section 5.5 we show

how the results on the time-scale $\tau_\hbar$ can be derived by using the considered in section 4 regular perturbation methods. In section 5.6 the example of "one-dimensional Einstein gas" is considered.

5.1 The Characteristic Times of Violation of Quasiclassical Approximation for Integrable Boson Systems

As mentioned in the Introduction the problem of comparing quantum dynamics with its corresponding classical motion arose with the creation of quantum mechanics. One of the first examples of this problem was discussed by Einstein [43] (see also the discussion on the "one-dimensional Einstein gas" in [44,45]). We return to this example in section 5.6.

A general approach to this problem can be formulated in the following way. Consider first an integrable boson system described by a Hamiltonian of the type (3.5) or (4.18). If $m + n \leq 2$, then the corresponding Heisenberg equations for the operators $\mathbf{a}^+(t)$ and $\mathbf{a}(t)$ are linear and quantum and classical dynamics for expectation values will actually coincide. The only difference will be the renormalization of some coefficients. This property is easily seen from equation (4.33). If $m + n \leq 2$ all quantum terms vanish in (4.33). So, the question arises: what parameters characterize a system whose quantum dynamics differs from the corresponding classical motion ? Equation (4.33) implies that Hamiltonian (4.18) for such a system should at least include nonlinear terms with $m + n > 2$. Thus, we need at least two dimensionless parameters: a classical parameter of nonlinearity $\bar{\mu}$ and a quantum parameter $\kappa = I/\hbar$ connected with the Planck constant. The next question is, how many degrees of freedom should be chosen ? The answer is rather evident in the context of the integrable systems, for which separation of variables uncouples the original system into independent quantum systems, each with one degree of freedom. Thus, in order to investigate the "$\tau_\hbar$-problem" in the integrable case all we need is a quantum nonlinear system ($\kappa \neq 0, \bar{\mu} \neq 0$) with one degree of freedom. Thus, our attention focusses on the quantum nonlinear oscillator, with Hamiltonian (3.18). In this section we consider quantum dynamics for the time-dependent expectation values of rather an arbitrary operator function $\mathbf{f}(t)$ for Hamiltonian (3.18) and then compare it with its corresponding classical dynamics. This procedure allows us to introduce the time-scale $\tau_{\hbar}$, at which the quasiclassical approximation for a quantum nonlinear oscillator is violated by the quantum dynamics. We begin by presenting the solution of the equation (3.19) for time-dependent expectation value $f(t)$ for a rather arbitrary operator $\mathbf{f}$ (see (3.6)), namely

$$\mathbf{f}_{k,l} \equiv \mathbf{f}_{k,l}(0) = (\mathbf{a}^{+})^{k}\mathbf{a}^{l}. \tag{5.1}$$

It is easy to check (see solution (3.29)) that the solution of (3.19) for the time-dependent expectation value $f_{k,l}(t)$ defined in (3.2) has the form

$$f_{k,l}(t) = \exp\left\{i(\omega + \mu\hbar)(k-l)t + i\left[k(k-1) - l(l-1)\right]\mu\hbar t\right\} \tag{5.2}$$
$$\times \exp\left\{|\alpha|^{2}\left(e^{2i(k-l)\mu\hbar t} - 1\right)\right\}(\alpha^{*})^{k}\alpha^{l},$$

with initial condition

$$f_{k,l}(0) = (\alpha^{*})^{k}\alpha^{l}. \tag{5.3}$$

Now we consider the following problem: what is the difference over time between the quantum solution (5.2) and the corresponding classical limit ?

We present here a definition of the difference $\delta(t)$ between these solutions based on the CS $|\alpha>$ and in the next section will investigate this difference using an additional averaging over the initial density matrix.

Classical limit for (5.2) corresponds to

$$\hbar \to 0, \qquad \hbar|\alpha|^{2} \to I = constant, \tag{5.4}$$

where I is a characteristic action of the system. In our case

$$I = \hbar < \alpha|\mathbf{a}^{+}\mathbf{a}|\alpha >= \hbar|\alpha|^{2} = constant. \tag{5.5}$$

So, using (5.4) we derive from (3.19) and (5.2) for the classical limit of $f_{k,l}^{(cl)}(t)$ the following equation

$$\dot{f}_{k,l}^{(cl)} = i(\omega + 2\mu\hbar|\alpha|^2)(\alpha^* \frac{\partial}{\partial \alpha^*} - \alpha \frac{\partial}{\partial \alpha}) f_{k,l}^{(cl)}. \tag{5.6}$$

The solution of the equation (5.6) is

$$f_{k,l}^{(cl)}(t) = e^{i(\omega+2\mu I)(k-l)t}(\alpha^*)^k \alpha^l. \tag{5.7}$$

The following function $\delta(t)$ characterises the difference between two solutions: the quantum expectation value (5.2) and the corresponding classical limit $f_{k,l}^{(cl)}(t)$ (5.7)

$$\delta(t) = \frac{|f_{k,l}(t) - f_{k,l}^{(cl)}(t)|}{|f_{k,l}^{(cl)}(t)|} = |\exp\{i\mu\hbar t(k-l) + i[k(k-1) - l(l-1)]\mu\hbar t \tag{5.8}$$
$$+|\alpha|^2 \left(e^{2i(k-l)\mu\hbar t} - 1\right) - 2i(k-l)\hbar|\alpha|^2 \mu t\} - 1|.$$

According to our definition

$$\delta(0) = \delta(t)|_{\hbar=0} = 0. \tag{5.9}$$

So, to estimate the characteristic time-interval $(0, \tau_\hbar)$ in which quantum dynamics is well approximated by the classical solution we need to find a region of parameters in (5.8) where $\delta(t) \ll 1$. This may be done in the following way. Define $\tau_\hbar$ as a solution of the equation

$$\delta(\tau_\hbar) = constant. \tag{5.10}$$

Since the function $\delta(t)$ in (5.8) is quasiperiodic there may exist an infinite number of solutions of equation (5.10). We deal only with the solution $\tau_\hbar = min(\tau_\hbar)$. For most problems the time-scale $\tau_\hbar$ must be found by numerical calculations. Usually, in numerical experiments (see below) we choose in (5.10): $constant = 10^{-2}$ or $2 \cdot 10^{-2}$ ("1% or 2%-criterion" [20]). For the case (5.8) we may estimate the time-scale $\tau_\hbar$ analytically. Introduce the following dimensionless parameter

$$\bar{\epsilon} = \bar{\mu}\epsilon \equiv \frac{\bar{\mu}}{\kappa}, \quad (\bar{\mu} = \frac{\mu I}{\omega}, \quad \kappa = \frac{I}{\hbar}) \tag{5.11}$$

This parameter has a simple physical meaning. It characterizes the quantum effects depending simultaneously on the classical parameter of nonlinearity $\bar{\mu}$ and the quantum parameter ϵ. Also we use the dimensionless time $\tau = \omega t$. Then, one has from (5.8)

$$\delta(\tau) = |\exp\{i\bar{\epsilon}\tau(k-l) + i[k(k-1) - l(l-1)]\bar{\epsilon}\tau + \kappa\left(e^{2i(k-l)\bar{\epsilon}\tau} - 1\right) - 2i(k-l)\bar{\mu}\tau\} - 1|. \tag{5.12}$$

The function $\delta(\tau)$ is quasiperiodic with the following periods for given k and l

$$\tau_1 = \frac{\pi}{\bar{\epsilon}(k-l)}, \quad \tau_2 = \frac{2\pi}{\bar{\epsilon}[k(k-1) - l(l-1)]}, \quad \tau_3 = \frac{\pi}{\bar{\mu}(k-l)} \tag{5.13}$$

To estimate the time-scale $\tau_\hbar$ note that we are interested in the quasiclassical region of parameters where $\kappa \gg 1$. For the quantum parameter $\bar{\epsilon}$ we shall use the following inequality

$$\bar{\epsilon} = \frac{\bar{\mu}}{\kappa} \sim \hbar \ll 1, \tag{5.14}$$

and expand the expression (5.12) up to the first order in $\bar{\epsilon}$. We have from (5.12)

$$\delta(t) = |i(k-l)\tau + i[k(k-1) - l(l-1)]\tau - 2(k-l)^2\tau^2\bar{\mu}|\bar{\epsilon} + O(\bar{\epsilon}^2). \tag{5.15}$$

Consider now two different cases: a) $k \sim l$, and b) $k \gg l$, (or $k \ll l$).

Case a): Put, for example, $k = 0; l = 1$. Then, it follows from (5.15) that

$$\delta(\tau) = |\tau + 2i\tau^2\bar{\mu}|\bar{\epsilon} + O(\bar{\epsilon}^2), \quad (k = 0, l = 1). \tag{5.16}$$

So, in this case we have two characteristic times $\tau^{(1)}$ and $\tau^{(2)}$ when $\delta(\tau^{(1,2)}) \sim 1$

$$\tau^{(1)} = \frac{1}{\bar{\epsilon}}, \quad \tau^{(2)} = \frac{1}{\sqrt{\bar{\mu}\bar{\epsilon}}}, \tag{5.17}$$

and

$$\frac{\tau^{(2)}}{\tau^{(1)}} = \sqrt{\frac{\bar{\epsilon}}{\bar{\mu}}} = \sqrt{\epsilon} \ll 1. \tag{5.18}$$

The inequality (5.18) means that in the case a) the characteristic time-scale $\tau_\hbar$ should be chosen as

$$\tau_\hbar = \tau^{(2)} = \frac{\sqrt{\kappa}}{\bar{\mu}}, \quad (k = 0, l = 1). \tag{5.19}$$

This time-scale $\tau_\hbar$ actually coincides with that found in [2].

Case b): Put $k \gg l$. Then, we have from (5.15)

$$\delta(\tau) \approx |1 + 2i\tau\bar{\mu}|\bar{\epsilon}\tau k^2 + O(\bar{\epsilon}^2). \tag{5.20}$$

Again, we have two characteristic times $\tau^{(1)}$ and $\tau^{(2)}$ when $\delta(\tau^{(1,2)}) \sim 1$

$$\tau^{(1)} = \frac{1}{\bar{\epsilon}k^2}, \quad \tau^{(2)} = \frac{1}{k\sqrt{\bar{\epsilon}\bar{\mu}}}, \tag{5.21}$$

and one should expect the violation of quasiclassical approach at these times. Assume that the following condition is satisfied

$$\frac{\tau^{(2)}}{\tau^{(1)}} = k\sqrt{\frac{\bar{\epsilon}}{\bar{\mu}}} \sim \hbar \ll 1. \tag{5.22}$$

Then, we have in the case b)

$$\tau_\hbar = \tau^{(2)} = \frac{\sqrt{\kappa}}{k\bar{\mu}}, \quad (k \gg l). \tag{5.23}$$

The same result will be when $l \gg k$ with the substitution $k \leftrightarrow l$.

The result (5.23) means that even for such a simple quantum nonlinear system as (3.18) one cannot introduce a universal time-scale $\tau_\hbar$, which depends only on parameters of Hamiltonian such as $\bar{\mu}$ and κ. In the general case, as it is seen from (5.23), this time-scale $\tau_\hbar$ depends also on the chosen operator $\mathbf{f}$ (5.1) whose expectation value we measure. As follows from (5.23) the quantum dynamics of expectation values for large momenta ($k \gg l$ or $l \gg k$) differs from the classical dynamics for rather shorter times, proportional to the factor $\sim 1/k$.

We also present results for the dynamical evolution of the expectation value for operator

$$\mathbf{f} = e^{x\mathbf{a}^+} e^{y\mathbf{a}}, \tag{5.24}$$

which appears in various problems (x, y are parameters). For this, we should solve the equation for $f(t)$ (3.19) with initial condition

$$f(0) =< \alpha | e^{x\mathbf{a}^+} e^{y\mathbf{a}} | \alpha >= e^{x\alpha^*} e^{y\alpha}.$$

The solution has the form

$$f(t) = \sum_{k=0}^{\infty} \sum_{l=0}^{\infty} \frac{x^k y^l}{k!l!} f_{k,l}(t), \tag{5.25}$$

where the c-number function

$$f_{k,l}(t) =< \alpha | \left(\mathbf{a}^+(t) \right)^k (\mathbf{a}(t))^l | \alpha > \tag{5.26}$$

is given by the expression (5.2). The classical limit for (5.25) is

$$f_{cl} = \exp\{x e^{i(\omega+2\mu I)t} \alpha_0^*\} \exp\{y e^{-i(\omega+2\mu I)t} \alpha_0\}. \tag{5.27}$$

To find the time-scale $\tau_\hbar$ we again estimate the function

$$\delta(t) = |f(t) - f^{(cl)}(t)| / |f^{(cl)}(t)|, \tag{5.28}$$

and find the solution of equation (5.10). Formally, expression (5.25) includes large momenta with $k \gg l$ and $l \gg k$, and according to the estimate (5.23) these terms have rather small time-scale $\tau_\hbar$ of violation of quasiclassical limit. Nevertheless, the contribution of these large momenta in (5.25) is small because of the $k!$ and $l!$ factors. In this case, the time-scale $\tau_\hbar$ of violation of the quasiclassical solution (5.25) can be estimated in the same way as above, and has an order (5.19).

5.2 Violation of Quasiclassical Approximation Upon Taking into Account an Additional Averaging over the Initial Density Matrix

As is known, in CS $|\alpha >$ the uncertainty of the wave packet at $t = 0$ is of the order $\hbar$ (see property 1 of CSs in section 2.2). Therefore, in the classical limit ($\hbar \to 0$) classical solutions (5.7), (5.27) correspond to the dynamics of an individual classical trajectory. In this case the corresponding quantum expectation values $f_{k,l}(t)$ in (5.2) and $f(t)$ in (5.26) are reduced to the evolution of one point in the classical phase space. So, the problem arises of deriving the time-scale $\tau_\hbar$ using an additional averaging of an arbitrary operator $\mathbf{f}(t)$ over the initial density matrix. For this, we introduce at $t = 0$ a density matrix $\hat{\rho}_0$, represented it in the form [30,31]

$$\hat{\rho}_0 = \frac{1}{\pi}\int d^2\alpha P(\alpha^*,\alpha)|\alpha><\alpha|, \tag{5.29}$$

where the weight function $P(\alpha^*,\alpha)$ is normalized according to

$$Tr(\hat{\rho}_0) = \frac{1}{\pi}\int d^2\alpha P(\alpha^*,\alpha) = 1. \tag{5.30}$$

The integration in (5.29) is defined in the following way

$$d^2\alpha \equiv dxdy = \frac{1}{2\hbar}dI\,d\theta, \quad \alpha = x + iy = \sqrt{\frac{I}{\hbar}}e^{-i\theta}, \tag{5.31}$$

$$(-\infty < x, y < \infty, \quad 0 \le \theta \le 2\pi, \quad 0 \le I < \infty).$$

To calculate the expectation value of the operator $\mathbf{f}(t)$ in (3.6) we have the expression

$$\begin{aligned}\bar{f}(t) = Tr\left(\hat{\rho}_0\mathbf{f}(t)\right) &= \frac{1}{\pi^2}\int\int d^2\alpha d^2\beta P(\alpha^*,\alpha) <\beta|\alpha><\alpha|\mathbf{f}(t)|\beta> \\ &= \frac{1}{\pi}\int d^2\alpha P(\alpha^*,\alpha) f(\alpha^*,\alpha,t),\end{aligned} \tag{5.32}$$

where the integration over β is made explicitly using the formula (2.30), and $f(\alpha^*,\alpha,t)$ is a diagonal matrix element in CS $|\alpha>$

$$f(\alpha^*,\alpha,t) \equiv <\alpha|\mathbf{f}(t)|\alpha>. \tag{5.33}$$

Expression (5.32) implies that to calculate the expectation value $\bar{f}(t)$ one needs to know only the diagonal matrix element (5.33) of the operator $\mathbf{f}(t)$. This circumstance plays a rather important role in simplifying the theory and justifies the use of equations (3.11),

(3.12) written only for the diagonal matrix element of the operator $\mathbf{f}(t)$ in the CSs.

Note, that the density matrix $\hat{\rho}_0$ in (5.29) can be easily expressed in the $|n>$- representation by matrix elements

$$(\hat{\rho}_0)_{n,m} \equiv < n|\hat{\rho}_0|m> = \frac{1}{\pi}\int d^2\alpha P(\alpha^*,\alpha)e^{-|\alpha|^2}\frac{\alpha^n(\alpha^*)^m}{\sqrt{n!m!}}, \quad (5.34)$$

where formula (2.30) was used. If density matrix $\hat{\rho}_0$ is given initially in the $|n>$-representation by matrix elements $(\hat{\rho}_0)_{n,m}$, then, to find the weight function $P(\alpha^*,\alpha)$ one should consider (5.34) as an integral equation for $P(\alpha^*,\alpha)$.

For example, we could choose the weight function $P(\alpha^*,\alpha)$ in the form

$$P(\alpha^*,\alpha) = \nu\exp(-\nu|\alpha-\alpha_0|^2). \quad (5.35)$$

In (5.35) ν and α_0 are parameters characterizing the width and the center of the initial distribution. We consider some properties of the weight function (5.35). Introduce at $t = 0$ the operators of "momentum" $\mathbf{p}$ and "coordinate" $\mathbf{q}$

$$\mathbf{p} = i\sqrt{\frac{\hbar}{2}}(\mathbf{a}^+ - \mathbf{a}), \quad \mathbf{q} = \sqrt{\frac{\hbar}{2}}(\mathbf{a}^+ + \mathbf{a}), \quad [\mathbf{q},\mathbf{p}] = i\hbar. \quad (5.36)$$

Using (5.32), (5.35), (5.36) we find for the uncertainty condition (2.4)

$$\sigma_p\sigma_q \equiv \sqrt{\overline{(\mathbf{p}-\bar{p})^2}\cdot\overline{(\mathbf{q}-\bar{q})^2}} = \frac{\hbar}{2}(1+\frac{2}{\nu}). \quad (5.37)$$

It follows from the definition (5.35) that in the classical limit ($\hbar \to 0$)

$$\lim_{\hbar\to 0}\frac{\nu}{\hbar} = \nu_0 = constant. \quad (5.38)$$

So the value $\sigma_p\sigma_q$ in (5.37) does not go to zero as $\hbar \to 0$ (as it occurs in CS $|\alpha>$)

$$\lim_{\hbar\to 0}\sigma_p\sigma_q = \frac{1}{\nu_0}.$$

Next we calculate the time-dependance of the expectation value of the operator

$$\mathbf{f}_{k,l}(t) = \left(\mathbf{a}^+(t)\right)^k((\mathbf{a}(t))^l, \quad (5.39)$$

using an additional averaging over the density matrix $\hat{\rho}_0$ (5.29), (5.35). Let us choose the simplest case $k = 0, l = 1$. Then, we have from (5.2) for the time-evolution of the expectation value of the operator $\mathbf{f}_{0.1}(t) \equiv \mathbf{a}(t)$ in CS $|\alpha>$

$$\alpha(t) \equiv <\alpha|\mathbf{a}(t)|\alpha> = e^{-i(\omega+\mu t)t} \exp\left\{|\alpha|^2 \left(e^{-2i\mu\hbar t} - 1\right)\right\} \cdot \alpha. \quad (5.40)$$

Using (5.29), (5.32), (5.35) and (5.40) yields

$$\bar{\alpha}(t) \equiv Tr\left(\hat{\rho}_0 \mathbf{a}(t)\right) \quad (5.41)$$

$$= \frac{\nu}{\pi} \int d^2\alpha e^{-\nu|\alpha-\alpha_0|^2 - i(\omega+\mu\hbar)t} \exp\left\{|\alpha|^2 \left(e^{-2i\mu\hbar t} - 1\right)\right\} \cdot \alpha$$

$$= \frac{\bar{\alpha}\nu_0^2}{[\nu_0 - \eta(t)]^2} \exp\left\{-i\left(\omega + \mu\hbar\right)t + \frac{I_0\nu_0\eta(t)}{[\nu_0 - \eta(t)]}\right\},$$

where

$$\eta(t) = \frac{1}{\hbar}\left(e^{-2i\mu\hbar t} - 1\right); \quad (5.42)$$

$$\alpha_0 \equiv a + ib, \quad I_0 \equiv \hbar(a^2 + b^2).$$

In the classical limit we find for $\bar{\alpha}_{cl}(t)$

$$\bar{\alpha}_{cl}(t) = \frac{\alpha_0 \nu_0^2 e^{-\frac{2i\nu_0 I_0 \mu t}{(\nu_0 + 2i\mu t)}}}{(\nu_0 + 2i\mu t)^2}. \quad (5.43)$$

We introduce as before the function $\delta(t)$

$$\delta(t) = |\bar{\alpha}(t) - \bar{\alpha}_{cl}(t)| / |\bar{\alpha}_{cl}(t)|, \quad (5.44)$$

which characterizes the difference between quantum and corresponding classical solutions. To estimate the time-scale $\tau_\hbar$ of violation of quasiclassical approximation we expand the quantum solution $\bar{\alpha}(t)$ in (5.41) in power of $\hbar$

$$\bar{\alpha}(t) = \bar{\alpha}_{cl}(t) + \left\{\frac{2\eta'_\hbar(\hbar = 0)}{[\nu_0 - \eta_{cl}(t)]} \right. \quad (5.45)$$

$$\left. -i\mu t + \frac{\nu_0^2 I_0 \eta'_{\hbar=0}(\hbar = 0)}{[\nu_0 - \eta_{cl}(t)]^2}\right\} \bar{\alpha}_{cl}(t)\hbar + O(\hbar^2),$$

where

$$\eta(t) = -2i\mu t - 2\mu^2 t^2 \hbar + O(\hbar^2), \tag{5.46}$$

$$\eta_{cl}(t) = -2i\mu t, \quad \eta'_{\hbar}(\hbar = 0) = -2\mu^2 t^2.$$

Using (5.43) and (5.45) we have from (5.44)

$$\delta(t) = \left| \frac{4\mu^2 t^2}{(\nu_0 + 2i\mu t)} - i\mu t - \frac{2\nu_0^2 I_0 \mu^2 t^2}{(\nu_0 + 2i\mu t)^2} \right| \hbar + O(\hbar^2). \tag{5.47}$$

Consider the time-interval

$$\frac{\nu_0}{\mu} \ll t \ll \frac{1}{\mu\hbar}. \tag{5.48}$$

For this time-interval it follows from (5.47) that

$$\delta(t) \approx \left| i\mu t + \frac{\nu_0^2 I_0}{2} \right| \hbar + O(\hbar^2). \tag{5.49}$$

Assume now that the following condition is satisfied

$$\nu_0^2 I_0 \hbar \sim 1, \tag{5.50}$$

which under the quasiclassical condition $\kappa \gg 1$ can be written in the form

$$\frac{\Delta I_0}{I_0} \sim \sqrt{\frac{\hbar}{I_0}} = \frac{1}{\sqrt{\kappa}} \ll 1, \tag{5.51}$$

where $\Delta I_0 \equiv 1/\nu_0$ is the characteristic width of the initial distribution (5.35) in the classical action. Then, the characteristic dimensionless time-scale $\tau_\hbar$ can be estimated from (5.49) to be

$$\tau_\hbar \sim \frac{\nu_0 \omega}{\mu} = \frac{1}{\bar{\mu}} \sqrt{\frac{I_0}{\hbar}} = \frac{\sqrt{\kappa}}{\bar{\mu}}, \tag{5.52}$$

$$(\bar{\mu} = \mu I_0/\omega, \quad \kappa = I_0/\hbar, \quad \tau = \omega t),$$

which coincides with the estimation (5.19) derived in CS $|\alpha>$ without additional averaging over the initial density matrix.

The result (5.52) needs some discussion. First, we note that according to (5.38) ν_0 is a classical parameter which does not depend on the Planck constant $\hbar$. Nevertheless, this circumstance is not in contradiction with the expression (5.50), where ν_0 is just expressed in terms of Planck constant $\hbar$. Instead, the result (5.52) means that already at the "classical" time-scale

$$\tau_\hbar = t\omega \sim \frac{\nu_0 \omega}{\mu}, \tag{5.53}$$

which formally includes only classical parameters, the quantum effects are significant, since $\delta(\tau_\hbar) \sim 1$. But, because of condition (5.50) (or (5.51)), the time-scale $\tau_\hbar$ can be expressed through Planck constant $\hbar$ (see (5.52)), and the numerical values of $\tau_\hbar$ in (5.19) and (5.52) are of the same order.

The estimate (5.52) is derived for a narrow distribution of the weight function (see (5.51)). For a rather wide distribution $\delta I \sim I_0$ the dimensionless time-scale $\tau_\hbar$ derived from (5.47) is of the order

$$\tau_\hbar \sim \kappa / \bar{\mu},$$

which is much larger than (5.52).

5.3 Time-Behavior of Quantum Correlation Functions

Consider now the time-behaviour of quantum correlation functions (QCF) of the type (3.46) in CSs and compare them with the corresponding classical limit. We present a simple example using Hamiltonian (3.18). As operators $\mathbf{F}$ and $\mathbf{G}$ in (3.57) we choose the following

$$\mathbf{F}(t) \equiv \sqrt{\hbar}\mathbf{a}^+(t), \quad \mathbf{G}(t) \equiv \sqrt{\hbar}\mathbf{a}(t). \tag{5.54}$$

The quantum correlation function in this example is then

$$P(\tau) \equiv < \alpha_0 | \mathbf{a}^+(t)\mathbf{a}(t) | \alpha_0 > - < \alpha_0 | \mathbf{a}^+(t) | \alpha_0 >< \alpha_0 | \mathbf{a}(t) | \alpha_0 > . \tag{5.55}$$

Using the expression (5.2) for the expectation value $f_{k,l}$ with $k = 1, l = 1, k = 1, l = 0$, and $k = 0, l = 1$, leads from (5.55) to

$$P(\tau) = 1 - e^{-4\kappa \sin^2 \bar{\epsilon}\tau}, \tag{5.56}$$

$$(\kappa = I_0/\hbar, \quad I_0 = \hbar|\alpha_0|^2, \quad \bar{\mu} = \mu I_0/\omega, \quad \bar{\epsilon} = \bar{\mu}/\kappa, \quad \tau = \omega t).$$

Note, in the classical limit $\kappa\bar{\epsilon}^2 = 0$ and $P(\tau) = 0$ for any τ. Now we derive an upper bound for $P(\tau)$. Let $\bar{\epsilon}\tau \ll 1$. Then, as $\sin(\bar{\epsilon}\tau) \leq \bar{\epsilon}\tau$ when $\bar{\epsilon}\tau \ll 1$, we have from (5.56)

$$P(\tau) \leq 1 - e^{-4\kappa\bar{\epsilon}^2\tau^2} \leq 4\kappa\bar{\epsilon}^2\tau^2, \tag{5.57}$$

where we have used the inequality

$$\exp(-4\kappa\bar{\epsilon}^2\tau^2) \geq 4\kappa\bar{\epsilon}^2\tau^2.$$

From (5.57) we have an estimate for $\tau_\hbar$

$$\tau_\hbar \sim \frac{1}{\bar{\epsilon}\sqrt{\kappa}} = \frac{\sqrt{\kappa}}{\bar{\mu}}, \tag{5.58}$$

which coincides with (5.19) and (5.52).

5.4 The Characteristic Times of Violation of Quasiclassical Approximation for Integrable Quantum Nonlinear Spin Systems

In this section we consider a rather simple but characteristic quantum nonlinear spin system and estimate its time-scale $\tau_\hbar$ of violation of the quasiclassical approximation. We show this time-scale $\tau_\hbar$ actually coincides with the estimate (5.19), (5.52) derived for the nonlinear boson systems considered above. To begin, consider a quantum nonlinear spin system whose Hamiltonian is

$$\mathbf{H}_0 = D\hbar^2(\mathbf{S}^z)^2, \tag{5.59}$$

where D is a parameter of anisotropy and $\mathbf{S}$ is a dimensionless spin operator with commutation relations (2.49). It is known this system has an exact solution (see, for example [41]). We apply here the approach presented in section 3.5 for investigating the expectation value dynamics.

For Hamiltonian (5.59) the operator $\hat{K}$ in (3.39) has the form

$$\hat{K} = iD\hbar\left(1+|\xi|^2\right)^{-2S}\left[\left(\xi^*\frac{\partial}{\partial\xi^*}-S\right)^2-\left(\xi\frac{\partial}{\partial\xi}-S\right)^2\right]\left(1+|\xi|^2\right)^{2S}$$
$$\equiv \hat{K}_{cl}+\hbar\hat{K}_q, \tag{5.60}$$

where

$$\hat{K}_{cl} = iD\left[\frac{4S_0|\xi|^2}{(1+|\xi|^2)}-\hbar(2S-1)\right]\left(\xi^*\frac{\partial}{\partial\xi^*}-\xi\frac{\partial}{\partial\xi}\right), \tag{5.61}$$

and

$$\hat{K}_q = iD\left(\xi^{*2}\frac{\partial^2}{\partial\xi^{*2}}-\xi^2\frac{\partial^2}{\partial\xi^2}\right). \tag{5.62}$$

In (5.61) S_0 is a dimensional spin length

$$S_0 = \hbar S. \tag{5.63}$$

Equations (5.60)-(5.62) present the differential operator $\hat{K}$ as the sum of two operators as in section 4.1. The operator $\hat{K}_{cl}$ (5.61) involves only first order derivatives and describes the classical dynamics of the system with Hamiltonian (5.59). Quantum effects are described by the operator $\hat{K}_q$ in (5.62) which involves second order derivatives. As spin operators $\mathbf{f}(t)$ in (3.42) choose the following

$$\mathbf{J}^+(t) = \mathbf{S}^+(t)/S, \quad \mathbf{J}^-(t) = \mathbf{S}^-(t)/S, \quad \mathbf{J}^z(t) = \mathbf{S}^z(t)/S. \tag{5.64}$$

Then, the exact c-number equations for the time-dependent expectation values in spin CS $|\xi>$ (2.51), namely

$$J^+(t) = <\xi|\mathbf{J}^+(t)|\xi>, \quad J^-(t) = <\xi|\mathbf{J}^-(t)|\xi>,$$
$$J^z(t) = <\xi|\mathbf{J}^z(t)|\xi>, \tag{5.65}$$

have the form (see (3.56))

$$\frac{\partial J^+(t)}{\partial t} = \hat{K}J^+(t), \quad \frac{\partial J^-(t)}{\partial t} = \hat{K}J^-(t), \quad \frac{\partial J^z(t)}{\partial t} = \hat{K}J^z(t), \tag{5.66}$$

with the operator $\hat{K}$ defined in (5.60) and with the initial conditions

$$J^{+}(0)=\frac{2\xi^{*}}{1+|\xi|^{2}},\quad J^{-}(0)=\frac{2\xi}{1+|\xi|^{2}},\quad J^{z}(0)=-\frac{1-|\xi|^{2}}{1+|\xi|^{2}}. \quad (5.67)$$

The solution of equations (5.66) with initial conditions (5.67) has the following explicit form

$$J^{+}(t)\equiv<\xi|\mathbf{J}^{+}(t)|\xi> \quad (5.68)$$

$$=\frac{2\xi^{*}\exp\left[-i(2S-1)\hbar Dt\right]}{[1+|\xi|^{2}\exp(2i\hbar Dt)]}\left[\frac{1+|\xi|^{2}\exp(2i\hbar Dt)}{1+|\xi|^{2}}\right]^{2S},$$

$$J^{-}(t)=\left(J^{+}(t)\right)^{*},\quad J^{z}(t)=J^{z}(0)=-\frac{1-|\xi|^{2}}{1+|\xi|^{2}}.$$

In the classical limit

$$S\to\infty,\quad \hbar\to 0,\quad \hbar S=S_{0}=constant, \quad (5.69)$$

we derive from (5.68) the classical solution

$$J_{cl}^{+}(t)=\frac{2\xi}{1+|\xi|^{2}}\exp\left[-2iS_{0}Dt\frac{(1-|\xi|^{2})}{(1+|\xi|^{2})}\right], \quad (5.70)$$

$$J_{cl}^{-}(t)=\left(J_{cl}^{+}(t)\right)^{*},\quad J_{cl}^{z}(t)=J_{cl}^{z}(0)==-\frac{1-|\xi|^{2}}{1+|\xi|^{2}}\equiv J. \quad (5.71)$$

As is seen from (5.70), (5.71) the frequency of classical nonlinear oscillations is

$$\omega(J)=2S_{0}DJ,\quad J\epsilon[-1,1]. \quad (5.72)$$

We present also the quasiclassical solution which is the expansion of (5.68) up to the first order in $\hbar$

$$J_{qc}^{+}(t)=\frac{2\xi^{*}}{(1+|\xi|^{2})}e^{i\omega(J)t} \quad (5.73)$$

$$\times\left[1+iD\hbar t-\frac{2i\hbar Dt|\xi|^{2}}{(1+|\xi|^{2})}-\frac{4\hbar S_{0}D^{2}t^{2}|\xi|^{2}}{(1+|\xi|^{2})^{2}}\right]$$

$$= \frac{2\xi^*}{(1+|\xi|^2)} e^{i\tau} \left[1 - \frac{i\tau}{2S} - \frac{\tau^2(1-J^2)}{4SJ^2} \right],$$

where the dimensionless time $\tau = \omega(J)t$ has been introduced.

We estimate now the time-scale $\tau_\hbar$ of violation of the quasiclassical approximation (5.73). For this, we introduce the function $\delta(t)$ (in the same way, as done above for boson systems), which characterizes the difference between quantum solution (5.68) and its corresponding classical solution (5.70). Then, we derive the characteristic time-scale $\tau_\hbar$ as the solution of the equation $\delta(\tau_\hbar) \sim 1$. We have for $\delta(\tau)$

$$\delta(\tau) = \frac{|J^+(\tau) - J^+_{(cl)}(\tau)|}{|J^+_{(cl)}(\tau)|} \approx \left| \frac{i\tau}{2S} + \frac{\tau^2(1-J^2)}{4SJ^2} \right|. \qquad (5.74)$$

Finally, we have from (5.74) an estimate for $\tau_\hbar$ in the quasiclassical region $S \gg 1$

$$\tau_\hbar = \sqrt{S}, \quad (S \gg 1). \qquad (5.75)$$

Actually, the result (5.75) has the same dependence on the parameters $\bar{\mu}$ and κ as derived before for boson systems. Indeed, in the case (5.59) the classical dimensionless parameter of nonlinearity is of the order unity (see (5.72)), i.e.,

$$\bar{\mu} = \frac{J}{\omega(J)} \frac{d\omega(J)}{dJ} \sim 1,$$

since we did not consider in (5.59) any linear terms (say, $\hbar\Omega \mathbf{S}^z$). The quasiclassical parameter κ can be expressed in the form

$$\kappa = S = \frac{S_0}{\hbar}. \qquad (5.77)$$

Using the definitions (5.76) and (5.77) we have from (5.75) an estimate for $\tau_\hbar$ in the same form as (5.19).

The additional averaging of an arbitrary spin operator $\mathbf{f}(t)$ over the initial density matrix $\hat{\rho}_0$ could be done in the same way as for boson systems. We introduce for this a density matrix

$$\hat{\rho}_0 = \int d\mu_s(\xi) P(\xi^*, \xi) |\xi><\xi|, \qquad (5.78)$$

where the weight function $P(\xi^*, \xi)$ satisfies the normalization condition

$$\int d\mu_s(\xi) P(\xi^*, \xi) = 1. \tag{5.79}$$

Define the time-dependent expectation value of an arbitrary operator $\mathbf{f}(t)$ in (3.41) derived by averaging over the initial density matrix $\hat{\rho}_0$ (5.78) as

$$\bar{f}(t) = Tr(\hat{\rho}_0 \mathbf{f}(t)) = \int d\mu_s(\xi) P(\xi^*, \xi) f(\xi^*, \xi, t), \tag{5.80}$$

where $f(\xi^*, \xi, t)$ is defined in (3.42). The density matrix $\hat{\rho}_0$ (5.78) may be expressed in the $|S, M>$ – representation (see section 2.3). Then, we have for the matrix elements of the density matrix $\hat{\rho}_0$

$$\rho^{(s)}_{M,M'} \equiv < M, S|\hat{\rho}_0|S, M' > \tag{5.81}$$

$$= \int d\mu_s(\xi) P(\xi^*, \xi) < M, S|\xi >< \xi|S, M' >= \int d\mu_s(\xi) U_M(\xi) U^*_{M'}(\xi)$$

$$= \frac{2S(2S+1)}{\pi\sqrt{(S+M)!(S-M)!(S+M')!(S-M')!}} \int d^2\xi \frac{|\xi|^{2S} P(\xi^*, \xi)}{(1+|\xi|^2)^{2S+2}}.$$

This expression allows one to calculate the matrix elements of the density matrix $\hat{\rho}_0$ for any weight function $P(\xi^*, \xi)$. The weight function $P(\xi^*, \xi)$ analogous to (5.35) can be chosen, for example, in the form

$$P(\xi^*, \xi) = \frac{\nu}{(2S+1)} (1+|\xi|^2)^2 e^{-\nu|\xi-\xi_0|^2}, \tag{5.82}$$

where the parameters ν and $\xi_0 = a + ib$ characterize the width and the center of the initial distribution. Using (2.58) and (5.82) we derive from (5.78) as $\nu \to 0$

$$\lim_{\nu \to 0} \hat{\rho}_0 = |\xi_0 >< \xi_0|, \tag{5.83}$$

which means that in this case we return to the pure spin CS $|\xi_0 >$. So the expression (5.80) allows one (at least numerically) to derive the time-dependent expectation values of an arbitrary operator $\mathbf{f}(t)$ if the solution of the equation (3.50) is known. Note, that the integration in (5.80) could be done by substitution of variables, namely

$$\xi = \cot\frac{\theta}{2}e^{i\varphi}, \quad (0 \leq \theta \leq \pi, \quad 0 \leq \varphi \leq 2\pi). \tag{5.84}$$

This substitution is the well known stereographic projection from the north pole of a sphere of unit diameter onto the complex plane $\xi = x + iy$ [35,36]. Then the integral (5.80) takes the form

$$\bar{f}(t) = \frac{(2S+1)}{4\pi}\int_0^{\pi} d\theta \int_0^{2\pi} d\varphi \sin\theta P(\theta,\varphi) f(\theta,\varphi,t). \tag{5.85}$$

We present here an upper bound for the time-scale $\tau_\hbar$ for the quantum expectation values (5.68) calculated with the additional averaging over the initial density matrix $\hat{\rho}_0$. As follows from the exact quantum expression (5.68) the additional averaging of $J^+(t)$ (5.80) does not change the two characteristic periods of the function $J^+(t)$

$$t_1 = \frac{\pi}{S_0 D}, \quad t_2 = \frac{2\pi}{\hbar D}. \tag{5.86}$$

So, the function $\bar{J}^+(t)$ is quasiperiodic and can be expressed in the form

$$\bar{J}^+(t) = \int d\mu_s(\xi) P(\xi^*,\xi) J^+(\xi^*,\xi,t) = \sum_{m,n} \bar{J}^+_{m,n} \exp\left(im\omega_1 t + in\omega_2 t\right), \tag{5.87}$$

$$(\omega_1 = 2\pi/t_1, \quad \omega_2 = 2\pi/t_2).$$

The period t_1 is classical, and the period t_2 is purely quantum. Then we have upper bound for the dimensionless time-scale $\tau_\hbar$

$$\tau_\hbar = \omega(J) t_2 \sim S = \frac{S_0}{\hbar} = \kappa. \tag{5.88}$$

As mentioned for the spin Hamiltonian (5.59) the dimensionless classical parameter of nonlinearity $\bar{\mu} \sim 1$ (see (5.76)) and the time-scale $\tau_\hbar$ (5.88) actually coincides with the upper bound (5.53) of this time-scale for the boson Hamiltonian (3.18) considered above.

5.5 Application of the Dynamical Perturbation Theory in Deriving the Time-Scale $\tau_\hbar$

As shown in section 5.4 the time-scale $\tau_\hbar$ of violation of quasiclassical approach can be derived from exact quantum solutions by comparing them with the corresponding classical solutions. However, in the general case (for example, for nonintegrable quantum Hamiltonians) it is impossible to derive analytically the exact quantum solutions. Nevertheless, even in this case we can estimate the characteristic time-scale $\tau_\hbar$ using the quasiclassical dynamical perturbation theory discussed in section 4. This method also allows one to estimate the characteristic dynamical behavior during the time-interval $0 < \tau < \tau_\hbar$ of quantum correlation functions introduced in section 5.3.

Here we consider how the discussed above dynamical perturbation theory allows one to derive the time-scale $\tau_\hbar$ for the integrable Hamiltonians (3.18) and (5.59).

Consider first the time-dependent expectation value $\alpha(\alpha_0^*, \alpha_0, t)$ for the boson Hamiltonian (3.18). For this function we have two exact equations. One of them is the linear equation (3.19) for $f(t) \equiv \alpha(\alpha_0^*, \alpha_0, t)$, which is convenient to rewrite here in the form

$$\dot{\alpha} = i\left(\omega + \mu\hbar + 2\mu|\alpha_0|^2\right)\left(\alpha_0^* \frac{\partial}{\partial \alpha_0^*} - \alpha_0 \frac{\partial}{\partial \alpha_0}\right)\alpha \tag{5.89}$$

$$+ i\mu\hbar\left((\alpha_0^*)^2 \frac{\partial^2}{\partial (\alpha_0^*)^2} - \alpha_0^2 \frac{\partial^2}{\partial \alpha_0^2}\right)\alpha,$$

$$\alpha(t) = \alpha(\alpha_0^*, \alpha_0, t), \quad \alpha(0) = \alpha_0.$$

In (5.89) we use the definitions introduced in section 4.2. The second equation for the same expectation value $\alpha(t)$ is the nonlinear one (3.39). This equation in the quasiclassical limit ($\kappa = I_0/\hbar \gg 1$ has the form (4.35).

As the first step to illustrate the application of the quasiclassical perturbation theory developed in section 4.1, we use the linear equation (5.89). We write the operator $\hat{K}$ in the form (4.1), which for the case (5.89) is

$$\hat{K} = \hat{K}_{cl} + \hbar \hat{K}_q, \tag{5.90}$$

$$\hat{K}_{cl} == i\left(\omega + \mu\hbar + 2\mu|\alpha_0|^2\right)\left(\alpha_0^*\frac{\partial}{\partial\alpha_0^*} - \alpha_0\frac{\partial}{\partial\alpha_0}\right),$$

$$\hat{K}_q = i\mu\left(\alpha_0^*\frac{\partial}{\partial\alpha_0^*} - \alpha_0\frac{\partial}{\partial\alpha_0}\right) + i\mu\left((\alpha_0^*)^2\frac{\partial^2}{\partial(\alpha_0^*)^2} - \alpha_0^2\frac{\partial^2}{\partial\alpha_0^2}\right).$$

Also we present solution $\alpha(t)$ in the form (4.2)

$$\alpha(t) = \alpha_{cl}(t) + \hbar\alpha_q(t), \tag{5.91}$$

where the classical solution $\alpha_{cl}(t)$ satisfies the following equation

$$\dot{\alpha}_{cl}(t) = \hat{K}_{cl}\alpha_{cl}(t), \quad \alpha_{cl}(0) = \alpha_{cl}(\alpha_0^*, \alpha_0) = \alpha_0. \tag{5.92}$$

Then, the quantum correction $\alpha_q(t)$ may be approximated (to first order in parameter $1/\kappa$) as in equation (4.5), which in the case (5.89) may be written as follows

$$\dot{\alpha}_q(t) = \hat{K}_{cl}\alpha_q(t) + \hat{K}_q\alpha_{cl}(t), \quad \alpha_q(0) = 0. \tag{5.93}$$

For the classical solution $\alpha_{cl}(\alpha_0^*, \alpha_0, t)$ we have from (5.92)

$$\alpha_{cl}(t) = \alpha_{cl}(\alpha_0^*, \alpha_0, t) = \alpha_0 \exp[-i(\omega + 2|\alpha_0|^2)t]. \tag{5.94}$$

To calculate the quantum correction $\alpha_q(t)$ from (5.93) we use the expression (4.17) for the solution by characteristics where the function D in (4.6) in our case is the function $\hat{K}_q\alpha_{cl}(t)$ expressed first as

$$\hat{K}_q\alpha_{cl}(t) = (-i\mu\alpha_0 - 4\mu^2|\alpha_0|^2\alpha_0 t)\exp[-i(\omega + 2\mu|\alpha_0|^2 t]. \tag{5.95}$$

The right part in (5.95) should be expressed in terms of α_{cl} and t. Using (5.94) and (5.95) gives an expression for the function D in (4.17), namely

$$\hat{K}_q\alpha_{cl}(t) = -i\mu\alpha_{cl} - 4\mu^2|\alpha_{cl}|^2\alpha_{cl}t. \tag{5.96}$$

Finally, formula (4.17) implies the explicit form for the quantum correction

$$\alpha_q(t) = \int_0^t (-\mu\alpha_{cl} - 4\mu^2|\alpha_{cl}|^2\alpha_{cl}\tau)d\tau \tag{5.97}$$

$$= -i\mu\alpha_{cl}t - 2\mu^2|\alpha_{cl}|^2\alpha_{cl}t^2$$
$$= -(i\mu t + 2\mu^2|\alpha_0|^2t^2)\alpha_0 \exp[-i(\omega + 2\mu|\alpha_0|^2)t.$$

Thus, the classical solution $\alpha_{cl}(t)$ in (5.94) and the first order quantum correction $\alpha_q(t)$ (5.97) implies the quasiclassical solution of equation (5.89)

$$\alpha_{qc}(t) \equiv \alpha_{cl}(t) + \hbar\alpha_q(t), \tag{5.98}$$

which is easily seen to be the expansion of the exact solution of equation (5.89) (see (5.40), where a substitution is made to denote the initial conditions as: $\alpha^* \equiv \alpha_0^*, \alpha \equiv \alpha_0$)

$$\alpha(t) = e^{-i(\omega+\mu\hbar)t} \exp\left\{|\alpha_0|^2\left(e^{-2i\mu\hbar t} - 1\right)\right\} \cdot \alpha_0 \tag{5.99}$$

up to terms including $\hbar$. At the same time, the quasiclassical solution $\alpha_{qc}(t)$ is the exact solution of the quasiclassical nonlinear equation (4.35). From the quasiclassical solution (5.98) one easily derives the time-scale $\tau_\hbar$ (5.19) using the condition $\delta(\tau_\hbar) \sim 1$, where

$$\delta(\tau) = \frac{|\alpha_{qc}(\tau) - \alpha_{cl}(\tau)|}{|\alpha_{cl}(\tau)|} = \hbar\frac{|\alpha_q(\tau)|}{|\alpha_{cl}(\tau)|}. \tag{5.100}$$

The same procedure applied to the spin system (5.59) gives the first order quantum correction

$$J_q^+(t) = \frac{2\xi^*}{(1+|\xi|^2)}e^{i\omega(J)t}\left[iDt - \frac{2iDt|\xi|^2}{(1+|\xi|^2)} - \frac{4S_0|\xi|^2D^2t^2}{(1+|\xi|^2)^2}\right]. \tag{5.101}$$

Also it is easy to check that the quantum correction (5.101) is the first order expansion in $\hbar$ of the exact quantum solution (5.68). This quantum correction allows an estimate of the time-scale $\tau_\hbar$ using the same formula as (5.100).

So, in the general case the time-scale $\tau_\hbar$ can be estimated from the expression

$$\hbar|f_q(\tau_\hbar)|/|f_{cl}(\tau_\hbar)| = 1, \tag{5.102}$$

where $f_q(\tau_\hbar)$ is the first order quantum correction which can be derived by the regular methods of the perturbation theory suggested in section 4, and $f_{cl}(\tau_\hbar)$ is the corresponding classical solution. Also it is possible to use the quantum correction $f_q(\tau_\hbar)$ to estimate the time-scale $\tau_\hbar$ with an additional averaging over the initial density matrix. In this case the equation (5.102) transforms into

$$\hbar|\bar{f}_q(\tau_\hbar)|/|\bar{f}_{cl}(\tau_\hbar)| = 1. \tag{5.103}$$

To estimate the time-scale $\tau_\hbar$ in numerical experiments we shall use in sections 7, 8, 11-14 the formula (5.10) with *constant* $= 10^{-2}$ and $2 \cdot 10^{-2}$.

5.6 The Time-Scale $\tau_\hbar$ for the One-Dimensional Einstein Gas

As was mentioned already, one of the first examples where the "$\tau_\hbar$ – *problem*" is discussed, is the "one-dimensional Einstein gas" [43-45]. This system consists of a gas of free quantum boson particles with mass m moving in a one-dimensional potential well with infinite walls. The distance between walls is l (see Fig. 1). In this case we have the Schrödinger equation for one particle

$$i\hbar\frac{\partial\Psi(x,t)}{\partial t} = -\frac{\hbar^2}{2m}\frac{\partial^2\Psi(x,t)}{\partial x^2}, \quad \Psi(0,t) = \Psi(l,t) = 0. \tag{5.104}$$

So, the solution of the eigenvalue problem is the following

$$\Psi_n(x) = \sqrt{\frac{2}{l}}\sin\left(\frac{\pi n x}{l}\right), \quad 0 \le x \le l, \tag{5.105}$$

$$E_n = \frac{\pi^2 n^2 \hbar^2}{2ml^2}, \quad n = 1, 2, \ldots.$$

Actually this system represents a quantum nonlinear oscillator with the frequencies

$$\omega_{n,q} = \frac{E_n - E_q}{\hbar} = \frac{\pi^2\hbar}{2ml^2}(n^2 - q^2), \quad (n, q = 1, 2, \ldots). \tag{5.106}$$

In the classical case the action I and the frequency $\omega(I)$ of the particle are

$$I = \frac{1}{2\pi}\oint p dx = \frac{|p|l}{\pi}, \quad \left(|p| = \frac{\pi n\hbar}{l}\right), \tag{5.107}$$

$$E(I) = \frac{\pi^2 I^2}{2ml^2}, \quad \omega(I) = \frac{dE(I)}{dI} = \frac{\pi^2 I}{ml^2}.$$

We have in this case the two characteristic parameters mentioned above: $\bar{\mu}$ and κ

$$\bar{\mu} = \frac{I}{\omega}\frac{d\omega(I)}{dI} = 1, \qquad \kappa = \frac{I}{\hbar} = \frac{|p|l}{\pi\hbar}. \tag{5.108}$$

In (5.107), (5.108) $|p|$ is the modulus of the characteristic momentum of a particle. A detailed analysis of this system is given in [45]. We present here only the upper bound for the time-scale $\tau_\hbar$ which follows from (5.106)

$$\tau_\hbar \equiv t_\hbar \omega \sim \kappa, \tag{5.109}$$

which coincides with the upper bound (5.88) for a nonlinear spin system.

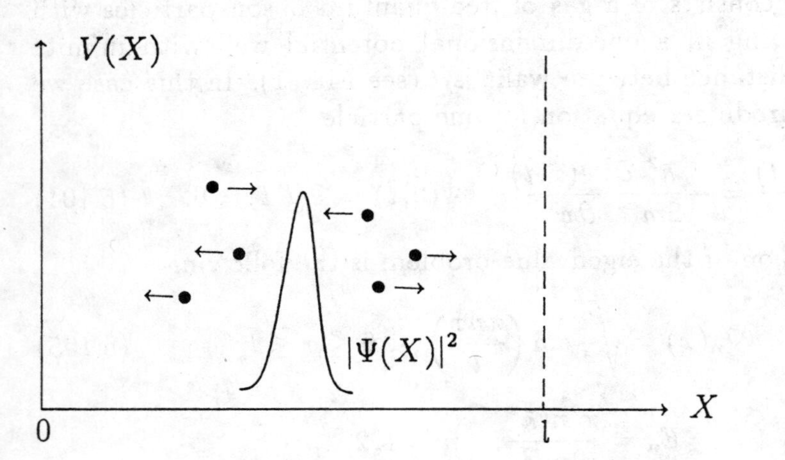

Fig. 1. The model of "one-dimensional Einstein gas"; boundary conditions for wave function: $\Psi(0,t) = \Psi(l,t) = 0$.

5.7 Conclusion

When considering the dynamical properties of quantum nonlinear boson and spin systems (systems which are described by Hamiltonians with non-equidistant energy spectrum) one of the main problems is connected with the estimation of the characteristic time-scale $\tau_\hbar$ when quantum dynamics differs significantly from the corresponding classical one. This problem is of special interest for nonintegrable quantum systems where in the corresponding classical limit one has the chaotic solutions. However, as the first step for understanding the peculiarities of the dependence of the time-scale $\tau_\hbar$ on the characteristic parameters, we have considered in this section the simpler, integrable quantum nonlinear systems. This section shows that rigorous perturbation methods may be used in considering the dynamics of quantum expectation values in the quasiclassical region of parameters and gives estimates of the time-scale $\tau_\hbar$ of violation of the quasiclassical solution. These methods are based on closed c-number equations for time-dependent quantum expectation values in boson and spin coherent states. It should be noted that: 1) There does not exist a universal time-scale $\tau_\hbar$ that depends only on the dimensionless parameters of the Hamiltonian, such as the classical parameter of nonlinearity $\bar{\mu}$ and the quasiclassical parameter κ (see (5.11)). In general case the time-scale $\tau_\hbar$ depends also on the expectation value one measures (see section 5.1, case b)) and for large momenta the value $\tau_\hbar$ may significantly decrease; 2) For smaller momenta a characteristic time-scale $\tau_\hbar$ can be introduced, and has the form

$$\tau_\hbar \sim \frac{\kappa^\alpha}{\bar{\mu}}, \tag{5.110}$$

where the typical values of α are: $\alpha = 1/2$ or 1. 3) Introducing an additional averaging of the time-dependent expectation values over the initial density matrix $\hat{\rho}_0$ can influence on the numerical value of $\tau_\hbar$ in a way that depends on the characteristic parameters of the initial distribution (for example, a width; see section 5.2).

To estimate the typical numerical value of $\tau_\hbar$ consider as an example the quantum nonlinear oscillator with the Hamiltonian (3.18). Section (5.2) shows in this case that the time-scale $\tau_\hbar$ can be expressed in terms of "classical" parameters: $\tau_\hbar = \tau_\hbar(\bar{\mu}, \Delta I)$, where ΔI is the width in action of the initial distribution of the density matrix $\hat{\rho}_0$ (see (5.51)). Then,

$$t_\hbar \equiv \tau_\hbar/\omega \sim I/\omega\bar{\mu}\Delta I, \qquad (5.111)$$

where I is a characteristic action of the oscillator and according to (5.52)

$$\frac{\Delta I}{\hbar} \sim \sqrt{\kappa} \gg 1, \quad (\kappa = I/\hbar). \qquad (5.112)$$

Choose the following parameters for the nonlinear quantum oscillator: the characteristic length of phase space $l \sim 10^{-5}cm$; the mass of the oscillator $m \sim 10^5 m_p \sim 10^{-19}g$; the velocity $v \sim 10^7 cm/sec$; the parameter of nonlinearity $\bar{\mu} \sim 10^{-1}$. Then, we have numerical estimates for the characteristic parameters

$$\kappa = I/\hbar \sim mvl/\hbar \sim 10^{10}, \qquad (5.113)$$

$$\Delta I/\hbar \sim \sqrt{\kappa} \sim 10^5, \quad \omega \sim v/l \sim 10^{12} sec^{-1},$$

$$E \sim \omega I + \mu I^2 \sim \hbar\omega(1+\bar{\mu})\kappa \sim 10^7 e.V., \quad \Delta E \equiv \hbar\omega \sim 10^{-3} e.V.,$$

$$t_\hbar \sim \frac{I}{\omega\bar{\mu}\Delta I} \sim 10^{-6} sec,$$

where E is a characteristic energy of the oscillator. Actually, in this case $\tau_\hbar$ also can be expressed in the form (5.52).

In the next section we shall apply the perturbation methods discussed above to derive the time-scale $\tau_\hbar$ for quantum nonintegrable boson and spin systems with 1.5 degrees of freedom. These systems are the simplest ones exhibiting dynamical chaotic behaviour in the corresponding classical limit.

6 Time-Scale $\tau_\hbar$ for a Kicked Quantum Nonlinear Oscillator

The following problem will be considered in sections 6-9. Consider a quantum nonintegrable Hamiltonian system. Nonintegrability means that the number of global independent integrals of motion is less then the number of degrees of freedom, and no separation of variables is possible neither in the Schrödinger equation, or in the corresponding classical equations of motion. In this case the classical dynamics is likely to be unstable in some regions of phase space, and leads to "dynamical chaos" [46-48]. Thus, the problem arises: what peculiarities of the dynamical and spectral properties of quantum systems arise if in the corresponding classical limit dynamical chaos takes place ? The field of scientific endeavor adressing this problem is called "quantum chaos" (see, for example, the monographs and reviews on this subject [4,5,13-15,19,49-54], and references therein). Our main interest focusses on the quasiclassical region of parameters where according to the correspondence principle at least for some finite interval of time τ: $0 < \tau < \tau_\hbar$, the quantum dynamics should approximately coincide with the classical chaotic motion.

In sections 6-9 we shall continue our discussions of the time-scale $\tau_\hbar$ of the applicability of the quasiclassical approximation, but for classically chaotic systems. For this we choose in these sections the simplest case - the case of quantum boson and spin systems with 1.5 degrees of freedom. In section 6 we investigate a system with a strong (global) classical chaos. This system models a nonlinear quantum oscillator kicked by a periodic sequence of δ-pulses. This system was the first case in which the time-scale $\tau_\hbar$ (1.2) was derived for classically chaotic motion [1]. In sections 7 and 8 we consider the dynamics of quantum nonlinear boson and spin systems with two interacting primary quantum nonlinear resonances. In the corresponding classical limit these systems exhibit globally chaotic behaviour in limited (bounded) regions of phase space. We calculate numerically $\tau_\hbar$ for these systems and show that the dependence $\tau_\hbar$ on the quasiclassical parameter is close to that in (1.2), under the condition of developed chaos in the classical limit. Finally, in section 9, we consider the dynamics of quantum linear oscillator kicked periodically in time by a sequence of δ-pulses. Under some resonant conditions this system realizes in the classical limit the so-called stochastic web [55,56]. We analyze quantum dynamics in this system in the regions of both strong and weak classical chaotic motion.

6.1 Quantum Map for C-Number Expectation Values

We consider here the problem of determining the character of the time-behavior of quantum corrections in the case when in the classical limit developed chaos takes place in an infinite region of phase space (unbounded classical diffusion in action) [1]. We use for this purpose the method discussed in section 4.2. As a model convenient for estimation of the time-scale $\tau_\hbar$ we choose the quantum system with Hamiltonian $\mathbf{H}_0$ (3.18) interacting via dipole coupling with an external field in the form of a time-periodic sequence of δ-pulses [1]

$$\mathbf{H} = \mathbf{H}_0 + \varepsilon \mathbf{V}(t), \quad ([\mathbf{a}, \mathbf{a}^+] = \hbar), \tag{6.1}$$

$$\mathbf{V}(t) = f(t)(\mathbf{a}^+ + \mathbf{a}), \quad f(t) = \sum_{n=-\infty}^{\infty} \delta(t - nT) = \frac{1}{T} \sum_{n=-\infty}^{\infty} \exp(in\nu t),$$

where ε and $T = 2\pi/\nu$ are the amplitude and the period of the external field. The Heisenberg equation of motion for the operator $\mathbf{a}$ has the form

$$i\hbar\dot{\mathbf{a}} = [\mathbf{a}, \mathbf{H}] = \hbar \Omega \mathbf{a} + \hbar\varepsilon f(t); \quad \Omega \equiv \Omega(\mathbf{a}^+\mathbf{a}) = \omega + \mu\hbar + 2\mu\mathbf{a}^+\mathbf{a}. \tag{6.2}$$

The evolution of the operator $\mathbf{a}(t)$ over the time-interval $(nT + 0)$ to $((n+1)T - 0)$ between two successive δ-pulses is determined by a phase rotation of the operators

$$\mathbf{a}_{n+1} = e^{-i\bar{\Omega}_n T} \bar{\mathbf{a}}_n, \quad \mathbf{a}^+_{n+1} = \bar{\mathbf{a}}^+ e^{i\bar{\Omega}_n T}, \tag{6.3}$$

where the following definitions are introduced

$$\mathbf{a}_n \equiv \mathbf{a}(nT - 0), \quad \bar{\mathbf{a}}_n \equiv \mathbf{a}(nT + 0), \tag{6.4}$$

$$\bar{\Omega}_n = \omega + \mu\hbar + 2\mu\bar{\mathbf{a}}^+_n \bar{\mathbf{a}}_n.$$

The variation of the operator $\mathbf{a}(t)$ under the perturbation $\mathbf{V}(t)$ at the moment $t = t_n = nT$ is determined by the shift

$$\bar{\mathbf{a}}_n = \mathbf{a}_n - i\varepsilon, \quad \bar{\mathbf{a}}^+_n = \mathbf{a}^+_n + i\varepsilon. \tag{6.5}$$

Equations (6.3) and (6.5) define a mapping $\hat{\mathbf{T}}$ for quantum noncommuting operators $\mathbf{a}_n^+$ and $\mathbf{a}_n$

$$\left(\mathbf{a}_{n+1}^+, \mathbf{a}_{n+1}\right) = \hat{\mathbf{T}}\left(\mathbf{a}_n^+, \mathbf{a}_n\right), \tag{6.6}$$

which will be investigated below according to the procedure analogous to that considered in section 4.2, but with appropriate modifications for the case of discrete transformations. We determine the quantum expectation values of the operators $\mathbf{a}_n$ and $\mathbf{a}_n^+$ in analogy with (4.21)

$$\alpha_n = < \alpha_0|\mathbf{a}_n|\alpha_0 > = \mathbf{a}_n^{(N)}(\alpha_0^*, \alpha_0); \tag{6.7}$$
$$\alpha_n^* = < \alpha_0|\mathbf{a}_n^+|\alpha_0 > = (\mathbf{a}_n^+)^{(N)}(\alpha_0^*, \alpha_0),$$

where n is a discrete time. Then, using the operator map (6.6) we get formally the c-number map for quantum expectation values (α_n^*, α_n)

$$\left(\alpha_{n+1}, \alpha_{n+1}^*\right) = \hat{\mathbf{G}}\left(\alpha_n, \alpha_n^*\right). \tag{6.8}$$

Since the map (6.6) is nonlinear, the explicit ordering of the operators $\mathbf{a}_n^+$ and $\mathbf{a}_n$ with respect to the initial operators $\mathbf{a}_0^+$ and $\mathbf{a}_0$ is possible to derive only in the quasiclassical approximation which is sufficient for our aims. From (6.3), (6.5), (6.7) we have

$$\alpha_{n+1} = e^{-i(\omega+\mu\hbar)t} < \alpha_0|e^{-2i\mu\bar{\mathbf{a}}_n^+\bar{\mathbf{a}}_n}\bar{\mathbf{a}}_n|\alpha_0 > . \tag{6.9}$$

First, we put the exponent in (6.9) into normal-ordered form. For this, we use the following formula

$$(\bar{\mathbf{a}}_n^+\bar{\mathbf{a}}_n)^p = \hat{N}\{(\bar{\mathbf{a}}_n^+\bar{\mathbf{a}}_n)^p + \hbar p(\bar{\mathbf{a}}_n^+\bar{\mathbf{a}}_n)^{p-1}\hat{D}(\bar{\mathbf{a}}_n^+, \bar{\mathbf{a}}_n) \tag{6.10}$$
$$+\frac{\hbar p(p-1)}{2}(\bar{\mathbf{a}}_n^+\bar{\mathbf{a}}_n)^{p-2}\hat{A}(\bar{\mathbf{a}}_n^+, \bar{\mathbf{a}}_n)\} + O(\hbar^2),$$

where the operator function $\hat{A}(\bar{\mathbf{a}}_n^+, \bar{\mathbf{a}}_n)$ is given by the formula

$$\hat{A}(\bar{\mathbf{a}}_n^+, \bar{\mathbf{a}}_n) = \hat{\mathbf{a}}_n^2 D(\bar{\mathbf{a}}_n^+, \bar{\mathbf{a}}_n^+)\hat{\mathbf{a}}_n^+\hat{\mathbf{a}}_n D(\bar{\mathbf{a}}_n^+, \bar{\mathbf{a}}_n^+) \tag{6.11}$$
$$+\hat{\mathbf{a}}_n^+\hat{\mathbf{a}}_n D(\bar{\mathbf{a}}_n, \bar{\mathbf{a}}_n^+) + (\hat{\mathbf{a}}_n^+)^2\hat{D}(\bar{\mathbf{a}}_n, \bar{\mathbf{a}}_n),$$

and the operator $\hat{D}(\mathbf{A},\mathbf{B})$ is defined in (4.29). The formula (6.10) can be derived in the same way as the formula (4.30). Using (6.10) it is possible to make the following normal ordering of the exponent

$$e^{s\bar{\mathbf{a}}_n^+\bar{\mathbf{a}}_n} = \sum_{p=0}^{\infty}(\bar{\mathbf{a}}_n^+\bar{\mathbf{a}}_n)^p = 1 + s\hat{N}\left\{\bar{\mathbf{a}}_n^+\bar{\mathbf{a}}_n + \hbar\hat{D}(\mathbf{a}_n^+,\mathbf{a}_n)\right\} \tag{6.12}$$

$$+\hat{N}\left\{\sum_{p=2}^{\infty}\frac{s^p}{p!}(\bar{\mathbf{a}}_n^+\bar{\mathbf{a}}_n)^p + \hbar p(\bar{\mathbf{a}}_n^+\bar{\mathbf{a}}_n)^{p-1}\hat{D}(\bar{\mathbf{a}}_n^+,\bar{\mathbf{a}}_n)\right\}$$

$$+\hat{N}\sum_{p=2}^{\infty}\frac{\hbar p(p-1)}{2}(\bar{\mathbf{a}}_n^+\bar{\mathbf{a}}_n)^{p-2}\hat{A}(\bar{\mathbf{a}}_n^+,\bar{\mathbf{a}}_n) + O(\hbar)$$

$$= \hat{N}\left\{\left[1 + \hbar s\hat{D}(\bar{\mathbf{a}}_n^+,\bar{\mathbf{a}}_n) + \frac{1}{2}\hbar s^2\hat{A}(\bar{\mathbf{a}}_n^+,\bar{\mathbf{a}}_n)\right]e^{s(\bar{\mathbf{a}}_n^+,\bar{\mathbf{a}}_n)}\right\} + O(\hbar^2).$$

Similarly, we can derive the formula

$$e^{s\bar{\mathbf{a}}_n^+\bar{\mathbf{a}}_n}\bar{\mathbf{a}}_n = \hat{N}\left\{e^{s\bar{\mathbf{a}}_n^+\bar{\mathbf{a}}_n}\bar{\mathbf{a}}_n + \hbar\hat{D}\left(e^{s\bar{\mathbf{a}}_n^+,\bar{\mathbf{a}}_n},\mathbf{a}_n\right)\right\} + O(\hbar^2) \tag{6.13}$$

$$= \hat{N}\{[\bar{\mathbf{a}}_n + 2\hbar\bar{\mathbf{a}}_n\hat{D}(\mathbf{a}_n^+,\mathbf{a}_n) + \hbar s\bar{\mathbf{a}}_n^+\hat{D}(\mathbf{a}_n,\mathbf{a}_n)$$

$$+\frac{1}{2}\hbar s^2\hat{\mathbf{a}}_n\hat{A}(\mathbf{a}_n^+,\mathbf{a}_n)]e^{s\bar{\mathbf{a}}_n^+\bar{\mathbf{a}}_n}\bar{\mathbf{a}}_n\} + O(\hbar^2).$$

Now all the essentials are in place for constructing the $\hat{G}$ map (6.8) in the quasiclassical approximation. The projection of (6.9) on the c-number phase space (α_n,α_n^*) is derived by replacing $\mathbf{a}_n \to \alpha_n$, $\mathbf{a}_n^+ \to \alpha_n^*$ in the right sides of (6.12) and (6.13)

$$\alpha_{n+1} = e^{-i(\omega+2\mu|\bar{\alpha}_n|^2)T} \tag{6.14}$$

$$\times\{[1 - i\mu\hbar T - 4i\mu\hbar T\hat{D}(\alpha_n^*,\alpha_n) - 2\hbar\mu^2T^2\hat{A}_n]\bar{\alpha}_n$$

$$-2i\mu\hbar T\bar{\alpha}_n^*\hat{D}(\alpha_n\alpha_n)\},$$

where function $\hat{A}_n$ is defined by the formula

$$\hat{A}_n = \bar{\alpha}_n^2\hat{D}(\alpha_n^*,\alpha_n^*) + (\bar{\alpha}_n^*)^2\hat{D}(\alpha_n,\alpha_n) \tag{6.15}$$

$$+|\bar{\alpha}_n|^2\left[\hat{D}(\alpha_n^*,\alpha_n) + \hat{D}(\alpha_n,\alpha_n^*)\right].$$

Equation (6.14) and its complex cojugate define the $\hat{G}$ map with quantum corrections of order $\hbar$.

6.2 Classical Limit

We present now the main results on the behavior of the system (6.1) in the classical limit [1] which will be used in the following. Formally the classical limit for the Hamiltonian (6.1) has the form

$$H(I,\theta,t) = \omega I + \mu I^2 + \frac{2\varepsilon\sqrt{I}}{T} \sum_{n=-\infty}^{\infty} \cos(\theta - n\nu t), \tag{6.16}$$

where the canonical substitution $\alpha = \sqrt{I}\exp(-i\theta)$ has been used. In the system with Hamiltonian (6.16) nonlinear resonances take place under the conditions

$$\Omega(I_n^0) \equiv \omega + 2\mu I_n^0 = n\nu. \tag{6.17}$$

Analogously to [57] we introduce the Hamiltonian of an isolated nonlinear resonance labeled by the number n

$$H_n = \mu J_n^2 + M_n \cos\Psi_n, \tag{6.18}$$

$$J_n \equiv I - I_n^0, \quad \Psi_n = \theta - n\nu t, \quad M_n = 2\varepsilon\sqrt{I_n^0}/T.$$

According to the Chirikov's criterion [57], the stochasticity in the behavior of the classical Hamiltonian system is connected with overlapping of nonlinear resonances of the type (6.18). In our case this criterion can be presented as follows

$$\bar{K}_n = \frac{\delta J_{n+1}/2 + \delta J_n/2}{\Delta I_n} > \bar{K}_n^{(cr)} \approx 1, \tag{6.19}$$

where δJ_n is a width in action of the nonlinear resonance (6.18)

$$\delta J_n = 4\sqrt{\varepsilon\sqrt{I_n^0}/\mu T}, \tag{6.20}$$

and ΔI_n is the distance between the neighboring nonlinear resonances in action

$$\Delta I_n = I_{n+1}^0 - I_n^0 = \nu/2\mu. \tag{6.21}$$

Thus, in the case (6.16) we have for $\bar{K}_n$

$$\bar{K}_n = \frac{2}{\pi}\sqrt{\varepsilon\mu T}\left[\left(I_{n+1}^0\right)^{1/4} + \left(I_n^0\right)^{1/4}\right]. \tag{6.22}$$

When $\bar{K}_n > 1$ the stochasticity takes place in most regions of the classical phase space (I, θ). When $\bar{K}_n \ll 1$ the motion is mainly regular. To analyze the conditions of stochasticity occuring in the system with Hamiltonian (6.16), a discrete classical map also can be used. This map can be derived from (6.15) with $\hbar = 0$ and has the form

$$\alpha_{n+1} = \exp\left\{-i\left(\omega + 2\mu|\alpha_n - i\varepsilon|^2\right)T\right\}(\alpha_n - i\varepsilon). \tag{6.23}$$

Using the substitution $\alpha_n = \sqrt{I_n}\exp(-i\theta_n)$ we derive from (6.23) a classical map for the system with the Hamiltonian (6.16) in the real variables

$$I_{n+1} = I_n + \varepsilon^2 + 2\varepsilon\sqrt{I_n}\sin\theta_n, \tag{6.24}$$

$$\theta_{n+1} = \arctan\left(\tan\theta_n + \frac{\varepsilon}{\sqrt{I_n}\cos\theta_n}\right) + \omega + 2\mu T I_{n+1}.$$

The variables (I_n, θ_n) are canonical, i.e.,

$$\mathrm{J} = \left|\frac{\partial(I_{n+1}, \theta_{n+1})}{\partial(I_n, \theta_n)}\right| = 1.$$

For the map (6.24) a stretching parameter for the phase θ (the parameter of stochasticity) $K_n = |\partial\theta_{n+1}/\partial\theta_n|$ can be introduced in the usual way (see, for example, [47]). In our case we have from (6.24)

$$\frac{\partial\theta_{n+1}}{\partial\theta_n} = \frac{1 + \varepsilon\sin\theta_n/\sqrt{I_n}}{1 + \varepsilon^2/I_n + 2\varepsilon\sin\theta_n/\sqrt{I_n}} + 4\varepsilon\mu T\sqrt{I_n}\cos\theta_n. \tag{6.25}$$

Under the condition of rather small classical perturbation $\varepsilon/\sqrt{I_n} \ll 1$, the parameter of stochasticity K_n has an order of magnitude

$$K_n \approx 4\varepsilon\mu T\sqrt{I_n^0} \approx (\frac{\pi}{2})^2\bar{K}_n^2, \tag{6.26}$$

where we put $I_{n+1}^0 \approx I_n^0$. As is seen from (6.26) the parameter of stochasticity K_n is connected with the Chirikov's overlapping parameter $\bar{K}_n$ as follows

$$K_n \approx \bar{K}_n^2, \tag{6.27}$$

which occurs also for other systems (see, for example, [47]). A characteristic stochastic trajectory generated by the map (6.24) is presented in Fig. 2a,b.

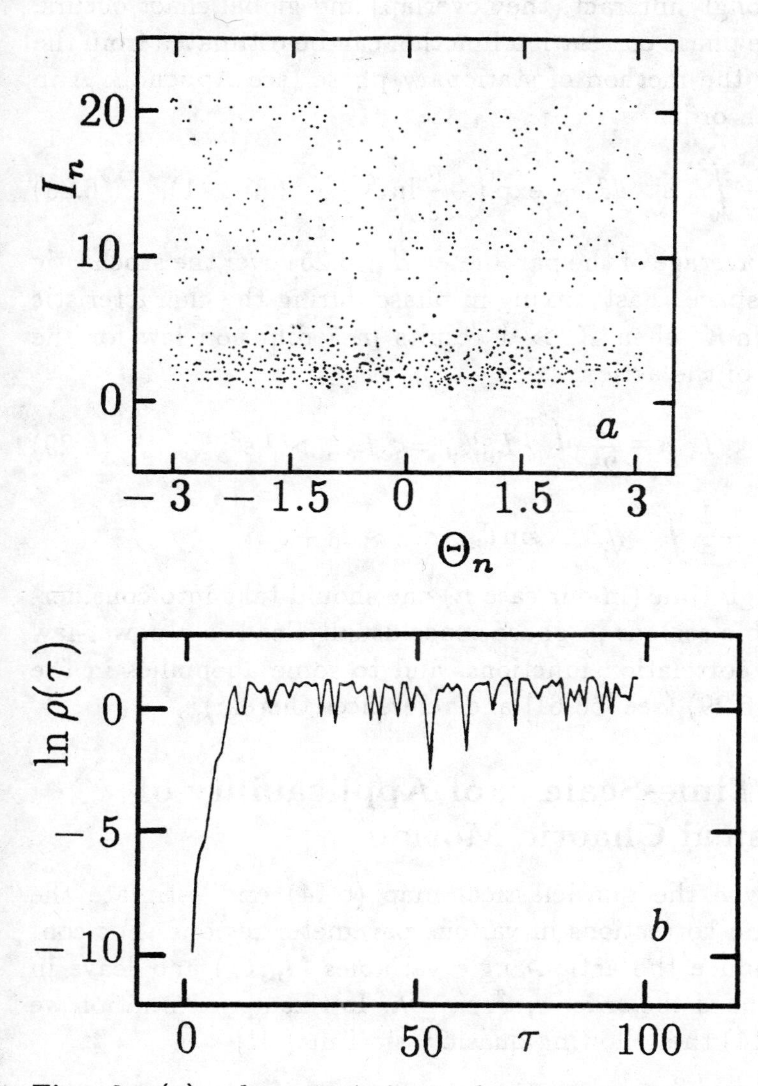

Fig. 2. (a): characteristic stochastic trajectory generated by the map (6.24); $\omega = 1; 2\mu T = 10; \varepsilon = 0.1; \theta_0 = 0; I_0 = 1$; (b): local instability of the stochastic trajectory shown in (a); $\rho(0) = 5 \cdot 10^{-5}$.

Thus, in the classical limit of $\hbar = 0$ the system (6.1) represents a system with an infinite number of interacting primary nonlinear resonances. In the region of phase space where $\bar{K}_n > 1$, the nonlinear resonances strongly interact (they overlap) and global chaos occurs. In this case the phase correlation function can be estimated from the map (6.24) by the method of stationary phase (see Appendix A in [6]), and has an order

$$R_c(n) = \frac{1}{2\pi}\int_0^{2\pi} e^{i\theta_n} d\theta_0 \sim \exp\left(-\frac{n}{2}\ln K\right), \quad (K \gg 1), \tag{6.28}$$

where K is an average of the parameter K_n (6.25) over the stochastic part of phase space. Fast mixing in phase during the characteristic time $\tau_c \sim 1/\ln K$ when $K \gg 1$ results in a diffusion law for the phase-average of the action

$$< I_n > \equiv \frac{1}{2\pi}\int_0^{2\pi} I_n d\theta_0 = < I_{n-1} > + \varepsilon^2 \tag{6.29}$$

$$+ \frac{1}{\pi}\int_0^{2\pi}\sqrt{I_{n-1}}\sin\theta_{n-1} d\theta_0 \sim I_0 + \varepsilon^2 n.$$

For large enough time (in our case n) one should take into consideration that stable regions in phase space usually lead to a power-law tail for phase correlation functions, and to some anomalies in the diffusion law (6.29) (see [58-61] and references therein).

6.3 The Time-Scale $\tau_\hbar$ of Applicability of Classical Chaotic Motion

Here we analyze the quasiclassical map (6.14) and estimate the role of quantum corrections in various parameter regions. For convenience introduce the action-angle variables (I_n, θ_n) and leave in (6.14) the terms of the order $\varepsilon, \varepsilon^2, \varepsilon\hbar, \varepsilon^2\hbar$. In this approximation we have from (6.14) the following quasiclassical map [1]

$$I_{n+1} = I_n + 2\varepsilon\sqrt{I_n}\sin\theta_n + \epsilon^2 - 4\hbar\beta_n\mu T I_n \tag{6.30}$$

$$\times \left\{\sin\theta_n + \cos\theta_n + \frac{\varepsilon}{\sqrt{I_n}}(2 + \cos 2\theta_n - \sin 2\theta_n - \cos\theta_n)\right\},$$

$$\theta_{n+1} = \theta_n + \omega T + \frac{\varepsilon}{\sqrt{I_n}} \cos\theta_n - \frac{\varepsilon^2}{2I_n} \sin 2\theta_n + 2\mu T I_{n+1}$$

$$+2\hbar\mu T \beta_n \left[1 + 2\frac{\varepsilon^2}{I_n}\left(1 + \sin^2\theta_n\right)\right].$$

In (6.30) the parameter β_n has the form

$$\beta_n = \frac{1}{4}\frac{I_n}{I_0}\left(\frac{\partial\theta_n}{\partial\theta_0}\right)^2. \tag{6.31}$$

Assume that developed chaos takes place in the classical limit. This means that for the parameter of stochasticity (6.26) the following condition is fulfilled

$$K = 4\varepsilon\mu T\sqrt{I} \gg 1, \tag{6.32}$$

where I is some characteristic action. Under the condition (6.32) the parameter β_n (6.31) may be estimated as

$$\beta_n \sim \frac{1}{4}\left(\frac{\partial\theta_n}{\partial\theta_{n-1}}\frac{\partial\theta_{n-1}}{\partial\theta_{n-2}}\cdots\frac{\partial\theta_1}{\partial\theta_0}\right)^2 \sim \frac{1}{4}K^{2n} = \frac{1}{4}\exp\left(2n\ln K\right).$$

Thus, the quantum corrections in the map (6.30) for the action I_n and phase θ_n have the order

$$\Delta I_q(n) \sim \hbar\mu T I \exp\left(2n\ln K\right), \tag{6.34}$$

$$\Delta\theta_q(n) \sim \hbar\mu T \exp\left(2n\ln K\right),$$

$$(\varepsilon/\sqrt{I} \ll 1, \quad K \gg 1).$$

Estimates (6.34) show that quantum corrections grow exponentially in time under the condition of the developed chaos in the classical limit. Let us estimate the characteristic time-scale $\tau_\hbar$ during which quantum corrections are small, and one may use the classical solutions under the condition of developed classical chaos. A rough estimate can be derived from the following conditions [1]

$$\Delta I_q(n) \ll I, \quad \Delta\theta_q(n) \ll 1. \tag{6.35}$$

These conditions give an estimate for the discrete time-scale $\tau_\hbar$ of the applicability of the classical approximation, in this case

$$\tau \ll \tau_\hbar = \frac{1}{2\ln K} \ln\left(\frac{1}{\zeta}\right), \tag{6.36}$$

$$(K \gg 1; \quad \zeta \equiv \mu\hbar T \ll 1).$$

Note, that the dimensionless quantum parameter ζ introduced above has the same physical meaming as the dimensionless quantum parameter $\bar{\epsilon}$ introduced in (5.11). The difference is that the parameter ζ is normalized by the period of the external field $T = 2\pi/\nu$, while the parameter $\bar{\epsilon}$ is normalized by the frequency of linear oscillations ω. Also we note that the expression for $\tau_\hbar$ given in formula (6.36) can be an over estimate because the quantum corrections in (6.30) are being compared with the smallest classical terms. For this reason we present a more general estimate of the time-scale $\tau_\hbar$ in the form

$$\tau_\hbar = \frac{1}{2\ln K} \ln\left(\frac{C}{\zeta}\right), \tag{6.37}$$

where C is a function that depends weakly on the classical parameters $C = C(\epsilon, \mu, K)$ (see the results of the numerical experiments below in this review). The estimate (6.36) can be expressed in terms of two characteristic dimensionless parameters: the classical parameter of stochasticity K (instead of the classical parameter of nonlinearity $\bar{\mu}$ for integrable systems), and the quantum (or quasiclassical) parameter κ (instead of the quantum (or quasiclassical) parameter $\kappa = I/\hbar$ for integrable systems). For simplicity we keep for the new quantum parameter the same notation κ. This parameter κ characterizes the number of quanta that an external field (interaction) gives to the oscillator during a single impulse. According to (6.24) we have for the parameter κ

$$\kappa = \frac{\Delta I}{\hbar} = \frac{\varepsilon\sqrt{I}}{\hbar}, \quad (\Delta I = I_{n+1} - I_n). \tag{6.38}$$

In the quasiclassical region $\kappa \gg 1$. Using (6.32), (6.38) we obtain from (6.36) an estimate of $\tau_\hbar$ expressed in terms of the two dimensionless parameters K and κ

$$\tau_\hbar \approx \frac{1}{2\ln K}\ln\left(\frac{\kappa}{K}\right). \tag{6.39}$$

Thus, the applicability of the classical approach when the classical limit exhibits global chaos is limited to the logarithmically small time interval $0 < \tau < \tau_\hbar \sim \ln(1/\hbar)$ (1.2) and not $\tau_\hbar \sim 1/\hbar^\alpha$ (1.1), as for integrable nonlinear quantum systems without unstable regions in phase space.

6.4 Quantum Boundary of Stochasticity

It follows from (6.39) that under the conditions

$$\kappa \gg K \gg 1 \tag{6.40}$$

the quantum dynamics nearly coincides with the classical chaotic motion during the finite time $0 < \tau < \tau_\hbar$. In this section we investigate this problem more accurately and derive the region of parameters K and κ in which over the finite times quantum dynamics can be considered as "shadowing" the chaotic motion of the corresponding classical system (6.16). For this, consider the expression for the expectation value of the operator action $\mathbf{I}_n = \hbar \mathbf{a}_n^+ \mathbf{a}_n$ in CS $|\alpha_0>$ for the system (6.2). We shall make an additional averaging over the initial ensemble with homogenous distribution over the initial phase of the oscillator θ_0

$$< I_{n+1} >= I_0 + \varepsilon^2(n+1) + i\varepsilon\sqrt{\hbar}\sum_{m=0}^{n}(< \alpha_m > - < \alpha_m^* >). \tag{6.41}$$

In (6.41)

$$< \cdots >= \frac{1}{2\pi}\int_0^{2\pi} d\theta_0 < \alpha_0| \cdots |\alpha_0 > . \tag{6.42}$$

In this section we use the dimensionless variable α_m ($[\mathbf{a}_m, \mathbf{a}_m^+] = 1$). The quantum expression (6.41) differs from the corresponding classical one only in the behavior of the "correlation functions" $< \alpha_m >$. In general the analysis of quantum correlation functions

$< \alpha_m >$ for an arbitrary time m is quite difficult. Nevertheless, for our intentions it is enough to consider only a few initial steps. From (6.9) we have for α_1 the exact result

$$\alpha_1 = \bar{\alpha}_0 \exp\left\{-i(\omega T + \zeta) + (e^{-i\zeta} - 1)|\bar{\alpha}_0|^2\right\}, \tag{6.43}$$

$$\bar{\alpha}_0 = \alpha_0 - i\varepsilon/\sqrt{\hbar}, \quad \zeta = \mu\hbar T.$$

Using (6.43) we get from (6.41)

$$< I_1 > /I_0 = 1 + \varepsilon^2/I_0, \tag{6.44}$$

$$< I_2 > /I_0 = 1 + 2\varepsilon^2/I_0 + R_1,$$

$$R_1 = \frac{4\varepsilon}{\sqrt{I_0}} \exp\left[-\frac{2(I_0+\varepsilon^2)\sin^2\zeta}{\hbar}\right]\left[-J_1(z)\sin\beta + \frac{\varepsilon}{\sqrt{I_0}} J_0(z)\cos\beta\right],$$

$$z = 2\varepsilon\sqrt{I_0}\sin 2\zeta\left(1 + \frac{i}{2}\tan\zeta\right)/\hbar,$$

$$\beta = \omega T + \zeta + \frac{(I_0+\varepsilon^2)}{\hbar}\sin^2\theta,$$

where $J_n(z)$ is the Bessel function. The function R_1 is the quantum correlation function. When $\zeta \ll 1$ expressions (6.43) and (6.44) are only slightly different from the corresponding classical limit ($\hbar = 0$)

$$R_1 = \frac{4\varepsilon}{\sqrt{I_o}}\{-J_1(K)\sin[\omega T + 2(I_0+\varepsilon^2)\mu T] \tag{6.45}$$

$$+ \frac{\varepsilon}{\sqrt{I_0}} J_0(K)[\omega T + 2(I_0+\varepsilon^2)\mu T]\}[1 + O(\zeta)].$$

Under the condition of stochasticity in the classical limit

$$K = 4\varepsilon\mu T\sqrt{I_0} \gg 1, \quad (\varepsilon/\sqrt{I_0} \ll 1), \tag{6.46}$$

we have from (6.45)

$$|R_1| \sim \frac{\varepsilon}{\sqrt{I_0 K}} \ll 1, \quad (\zeta \ll 1). \tag{6.47}$$

On the contrary, when $\zeta > 1$ it follows from (6.45) that the correlation function R_1 is purely quantum

$$R_1 \approx -\frac{4\varepsilon}{\sqrt{I_0}} \exp\left(-\frac{2I_0 \sin^2\zeta}{\hbar}\right) J_1(2\kappa), \tag{6.48}$$

$$(\kappa = \varepsilon\sqrt{I_0/\hbar}, \quad \zeta > 1).$$

As before, κ is about equal to the number of quanta that one impulse of the external field gives to the oscillator. From (6.47) and (6.48) it follows that the correlation function R_1 may be small in both cases $\zeta \ll 1, \zeta > 1$. Then, the quantum diffusion in the action during finite time is determined mainly by the same coefficient ε^2/I_0. However, a despite this numerical coincidence of the diffusion coefficients, the cases (6.47) and (6.48) are significantly different. Indeed, the motion of the quantum system in case (6.47) remains close during a finite time to the classical stochastic motion. The corresponding region of parameters can be written in the following way $\zeta \ll 1$, $K \gg 1$ or

$$\kappa \gg K \gg 1, \quad (K/\kappa = 4\zeta \ll 1). \tag{6.49}$$

The right inequality in (6.49) means that in the classical limit chaotic dynamics takes place in most regions of phase space. The left inequality corresponds to the condition

$$\zeta = \mu\hbar T = \frac{K}{4\kappa} \ll 1, \tag{6.50}$$

which means according to (6.36) the existence of the time-scale $\tau_\hbar > 0$ of the applicability of quasiclassical approximation. It is natural to call the condition

$$\kappa = K = 1 \tag{6.51}$$

a "quantum boundary of stochasticity" (QBS) [62]. Under the condition $\zeta > 1$ the diffusion in (6.45) is purely quantum, and can be distinguished from classical diffusion only by measuring quantum correlation functions such as R_1.

In connection with this QBS (6.51) note that the quasiclassical condition $\kappa \gg 1$ and the condition of stochasticity in the classical limit $K \gg 1$ are not sufficient for the existence of classical stochasticity over a finite time in the quantum system. The additional condition $\kappa \gg K$ (see (6.40)) is necessary to have the finite time $\tau_\hbar > 0$ (6.39). The inequality $\kappa \gg K$ has a simple physical meaning. During the time T between two neighboring impulses the cell in the phase space in action of the order $\Delta I \sim \hbar$ is spreading over the phase space in phase

$$\Delta\theta_\hbar \sim \frac{d\omega(I)}{dI}\Delta I \cdot T \sim \mu\hbar T = \zeta \sim K/\kappa, \quad (\omega(I) = \omega + 2\mu I). \tag{6.52}$$

Under the condition $\zeta \ll 1$ ($\kappa \gg K$) quantum effects of the phase spreading of the initial wave packet are small over a finite time. The condition $\zeta > 1$ ($\kappa < K$) means that during one period of the external field the initial wave packet of size $\hbar$ in action spreads significantly in phase, which leads to the differences between classical and quantum expectation values even during the time $t < T$.

Note that the QBS (6.51) derived above (which follows from the condition $\zeta \sim 1$) gives only a rough estimation. A more accurate estimate for the QBS has the form $\zeta \sim C(\varepsilon, \mu, K)$ where the function C includes the dependence on the main classical parameters of the system. An important problem in quantum chaos is the problem of deriving the time-scale $\tau_\hbar \sim \ln(1/\hbar)$ (1.2) under real experimental conditions. We discuss this subject in sections 11-14.

6.5 Additional Averaging over the Initial Density Matrix

We now make an additional averaging of the quantum expectation value α_n (6.15) over the initial density matrix (5.29), (5.35). Then we have

$$< \alpha_n > \equiv Tr\,(\mathbf{a}_n \hat{\rho}_0) = \int d^2\alpha_0 P(\alpha_0^*, \alpha_0)\alpha_n(\alpha_0^*, \alpha_0), \tag{6.53}$$

where the c-number function $\alpha_n(\alpha_0^*, \alpha_0)$ is determined in (6.15). We present the function $< \alpha_n >$ in (6.53) in the form

$$< \alpha_n > = < \alpha_n >_{cl} + \hbar < \alpha_n >_q + O(\hbar^2), \tag{6.54}$$

where $< \alpha_n >_{cl}$ is the classical part and $< \alpha_n >_q$ is the quantum correction. An estimate of the time when the quantum correction in (6.54) plays a significant role can be made from the condition

$$g = \hbar | < \alpha_n >_q / < \alpha_n >_{cl} | \sim 1. \quad (6.55)$$

A simple estimate of the order of magnitude in the case of the "narrow packet" ($\nu|\alpha_0|^2 \gg 1$) gives

$$g \sim \zeta \cdot \exp(2n \ln \tilde{K}), \quad (\tilde{K} = 4\varepsilon\mu T|\alpha_0|^2 \gg 1), \quad (6.56)$$

where ν and α_0 are parameters of the initial distribution (5.29), (5.35). From (6.56) we have an estimate for the discrete time-scale $n_\hbar$ of the applicability of quasiclassical approach, namely

$$n < n_\hbar \sim \frac{1}{2 \ln \tilde{K}} \ln \left(\frac{1}{\zeta} \right), \quad (6.57)$$

which coincides with (6.36).

7 The Time of Applicability of the Quasiclassical Approach in a Boson System of Two Interacting Quantum Nonlinear Resonances

As mentioned in section 6, when investigating dynamical chaos in classical Hamiltonian systems two convenient concepts are the "nonlinear resonance" (NR) and "interaction (overlapping) of NRs" [57]. Namely, the transition to chaos takes place under conditions of strong interaction of NRs in classical Hamiltonian systems [57]. These concepts also appear to be useful in the investigation of quantum chaos in Hamiltonian systems [63-66,9,15,53]. In this case quantum nonlinear resonances occur (QNR) [63,64]. Under the condition of strongly developed chaos in the classical limit, when an infinite number of primary NRs are overlapped (section 6), the time-scale $\tau_\hbar$ of applicability of the quasiclassical approach can be estimated analytically (see (6.36), (6.37), (6.39)), and is logarithmically small in terms of the quasiclassical parameter κ (6.38).

However, usually in physical systems the number of overlapping primary NRs is finite [53,66,67]. This leads to an increase in the measure of the stable component of motion in classical phase space, and also to the appearance of additional boundaries between "chaos" and "order". In this case, various anomaleous effects occur in the dynamical properties of the system [53,57-61,67]. As a result, analytical estimates of the time-scale $\tau_\hbar$ in this case in the deep quasiclassical region become significantly more difficult to obtain.

In this section we apply the method of dynamical perturbation theory discussed in section 4.2 which allows one to derive numerically the time-scale $\tau_\hbar$ for a system of two interacting QNRs in the deep quasiclassical region, and to show that the dependence of $\tau_\hbar$ on the quasiclassical parameter still has the logarithmic form (1.2) under the condition of developed chaos in the classical limit. We also show that in this case the quantum correlation functions grow exponentially in time. This effect can be verifyed in both numerical and real experiments.

7.1 The Main Equations

To model the interaction of two primary QNRs it is convenient to use the following Hamiltonian [64]

$$\mathbf{H} = \mathbf{H}_0 + \varepsilon f(t)\mathbf{V}, \tag{7.1}$$

$$\mathbf{H}_0 = \hbar\omega\mathbf{a}^+\mathbf{a} + \mu\hbar^2(\mathbf{a}^+\mathbf{a})^2,$$

$$\mathbf{V} = \sqrt{\hbar}(\mathbf{a}^+ + \mathbf{a}), \quad f(t) = \cos(\omega_1 t) + \cos(\omega_2 t),$$

$$\left([\mathbf{a}, \mathbf{a}^+] = 1, \quad \omega_2 > \omega_1\right),$$

which describes the dynamics of the quantum nonlinear oscillator (3.18) perturbed by an external field with two frequencies. In the classical limit the Hamiltonian (7.1) transforms into the following

$$H = H_0 + \varepsilon f(t)V, \quad H_0 = \omega I + \mu I^2, \quad V = 2\sqrt{I}\cos\theta, \tag{7.2}$$

after using the canonical transformation (3.20). In what follows we impose a condition of small nonlinearity,

$$\bar{\mu} = \frac{d\omega(I)}{dI}\frac{I}{\omega}|_{I=I_0} \sim \frac{\mu I_0}{\omega} \ll 1, \tag{7.3}$$

where $\omega(I) = \omega + 2\mu I$ and I_0 is a characteristic action of the system. We also assume the following resonant conditions to be satisfied

$$\omega(I_1) = \omega + 2\mu I_1 = \omega_1, \tag{7.4}$$

$$\omega(I_2) = \omega + 2\mu I_2 = \omega_2, \quad (I_2 > I_1).$$

The resonant conditions (7.4) allow one to exclude the fast motion. We introduce the sum and difference frequencies

$$\omega_0 = \frac{\omega_1 + \omega_2}{2}, \quad \nu_0 = \frac{\omega_2 - \omega_1}{2}, \tag{7.5}$$

where ω_0 is a fast frequency, and ν_0 is a slow one. Taking into account the definitions (7.5) allows one to express the Hamiltonian $\mathbf{H}$ (7.1) in the form

$$\mathbf{H} = \bar{\mathbf{H}}_0 + \bar{\mathbf{V}}, \tag{7.6}$$

$$\bar{\mathbf{H}}_0 = \hbar\omega_0 \mathbf{a}^+\mathbf{a},$$

$$\bar{\mathbf{V}} = \hbar(\omega - \omega_0)\mathbf{a}^+\mathbf{a} + \mu\hbar^2(\mathbf{a}^+\mathbf{a})^2 + 2\varepsilon\sqrt{\hbar}(\mathbf{a}^+ + \mathbf{a})\cos(\omega_0 t)\cos(\nu_0 t).$$

We then use the interaction representation with the Hamiltonian $\bar{\mathbf{H}}_0$. In this representation we have for Heisenberg operators $\mathbf{a}(t)$ and $\mathbf{a}^+(t)$

$$\mathbf{a}(t) = e^{i\bar{\mathbf{H}}_0 t/\hbar}\mathbf{a}e^{-i\bar{\mathbf{H}}_0 t/\hbar} = e^{-i\omega_0 t}\mathbf{a}, \tag{7.7}$$

$$\mathbf{a}^+(t) = e^{i\omega_0 t}\mathbf{a}^+.$$

Using (7.7) casts the Hamiltonian $\bar{\mathbf{V}}$ into the interaction representation

$$\bar{\mathbf{V}}_{int} = \hbar(\omega - \omega_0)\mathbf{a}^+\mathbf{a} + \mu\hbar^2(\mathbf{a}^+\mathbf{a})^2 + 2\varepsilon\sqrt{\hbar}\left(\mathbf{a}^+ e^{i\omega_0 t} + \mathbf{a}e^{-i\omega_0 t}\right) \tag{7.8}$$

$$\times \cos(\omega_0 t)\cos(\nu_0 t).$$

Taking together the resonant conditions (7.4), the condition of small nonlinearity (7.3) and also the condition of small interaction ($\varepsilon/\omega_0\sqrt{I} \ll 1$), allows us to exclude the fast motion with the frequency of the order $2\omega_0$ from the Hamiltonian (7.8). As a result we derive the quantum Hamiltonian of interaction of two QNRs which describes only the slow dynamics (see also [64])

$$\bar{\mathbf{V}}^{sl}_{int} = \hbar(\omega - \omega_0)\mathbf{a}^+\mathbf{a} + \mu\hbar^2(\mathbf{a}^+\mathbf{a})^2 + \varepsilon\sqrt{\hbar}(\mathbf{a}^+ + \mathbf{a})\cos(\nu_0 t). \quad (7.9)$$

Our further aim is to study quantum corections in the deep quasi-classical region ($I_0/\hbar \gg 1$) for the system (7.9). As in section 4.2 it is possible to derive for systems of the type (7.9) a closed c-number equation (4.33) in the quasiclassical approximation for expectation values, for example, for the function

$$\alpha(\alpha_0^*, \alpha_0, t) = < \alpha_0|\mathbf{a}|\alpha_0 > . \quad (7.10)$$

For Hamiltonian (7.9), equation (4.33) can be written in the form (see also (4.35))

$$i\dot{\beta} = [(\omega - \omega_0) + \mu\hbar + 2\mu|\beta|^2]\beta \quad (7.11)$$

$$+2\hbar\mu\frac{\partial\beta}{\partial\beta_0}\frac{\partial\beta}{\partial\beta_0^*}\beta^* + 4\hbar\mu\frac{\partial\beta^*}{\partial\beta_0}\frac{\partial\beta}{\partial\beta_0^*}\beta + \varepsilon\cos(\nu_0 t),$$

where $\beta = \alpha\sqrt{\hbar}$ is the renormalized expectation value. An approximate method of solving the equation (7.11) in quasiclassical region of parameters $|\beta|^2/\hbar = I/\hbar \gg 1$ consists in the following, which will also be used below in numerical simulations. In the quasiclassical region the quantum corrections we are interested in include the small parameter $\sim \hbar$. Consequently, the derivatives with respect to initial conditions in (7.11) can be well approximated by using the classical trajectories. For this porpose an additional approximate system of equations for these derivatives can be derived using (7.11)

$$i\frac{\partial}{\partial t}\left(\frac{\partial\beta}{\partial\beta_0}\right) = \left[(\omega - \omega_0) + 2\mu|\beta|^2\right]\frac{\partial\beta}{\partial\beta_0} + 2\mu\beta\left(\frac{\partial\beta}{\partial\beta_0}\beta^* + \frac{\partial\beta^*}{\partial\beta_0}\beta\right), \quad (7.12)$$

$$i\frac{\partial}{\partial t}\left(\frac{\partial\beta}{\partial\beta_0^*}\right) = \left[(\omega-\omega_0)+2\mu|\beta|^2\right]\frac{\partial\beta}{\partial\beta_0^*}+2\mu\beta\left(\frac{\partial\beta}{\partial\beta_0^*}\beta^*+\frac{\partial\beta^*}{\partial\beta_0^*}\beta\right).$$

Thus, the problem of deriving the solution of the equation (7.11) up to terms of order $\sim\hbar$ reduces to the problem of solving a system of three ordinary differential equations for the following functions

$$\beta,\ \frac{\partial\beta}{\partial\beta_0},\ \frac{\partial\beta}{\partial\beta_0^*}. \tag{7.13}$$

These equations are solved numerically in section 7.3.

7.2 Classical Limit

Before discussing the results of numerical simulation of equations (7.11), (7.12) we present in this section some characteristic properties of the behavior of the system (7.9) in the classical limit. In this case the Hamiltonian (7.9) takes the form

$$\bar{V}_{int}^{sl} = (\omega-\omega_0)I+\mu I^2+\varepsilon\sqrt{I}\left[\cos(\theta+\nu_0 t)+\cos(\theta-\nu_0 t)\right]. \tag{7.14}$$

The system with Hamiltonian (7.14) describes the interaction of two primary NRs. The centers of these NRs are given according to (7.4) and (7.5) by the following expressions

$$I_1=\frac{-\nu_0-(\omega-\omega_0)}{2\mu},\quad I_2=\frac{\nu_0-(\omega-\omega_0)}{2\mu}. \tag{7.15}$$

The distance in action between the centers of two NRs is

$$\Delta I\equiv I_2-I_1=\nu_0/\mu. \tag{7.16}$$

We estimate the width δI of each NR in (7.14) in the case of weak interaction between the NRs ($\bar{K}=\delta I/\Delta I\ll 1$). We have, for example, the following equations for the second NR in (7.14) (when the first NR is neglected)

$$\dot{I}=-\frac{\partial\bar{V}_{int}^{sl}}{\partial\theta}=\varepsilon\sqrt{I_0}\sin(\theta-\nu_0 t), \tag{7.17}$$

$$\dot{\theta} = \frac{\partial \bar{V}_{int}^{sl}}{\partial I} = \omega - \omega_0 + 2\mu I.$$

In (7.17) the value I_0 is

$$I_0 = \frac{I_1 + I_2}{2} = \frac{\omega_0 - \omega}{2\mu}, \tag{7.18}$$

and for simplicity we have assumed

$$\frac{I_2 - I_1}{I_1 + I_2} = \frac{2\nu_0}{\omega_0 - \omega} \ll 1. \tag{7.19}$$

Introducing in (7.17) the resonant phase $\Psi_2 = \theta - \nu_0 t$ and action $J_2 = I - I_2$ gives the resonant Hamiltonian for the second NR

$$H_r^{(2)} = \mu J_2^2 + \varepsilon\sqrt{I_0}\cos\Psi_2. \tag{7.20}$$

From (7.20) we may estimate the width in action for the second (and also the first) NR

$$\delta J_2 = \delta J_1 = \delta J = 2\sqrt{2\varepsilon\sqrt{I_0}/\mu}. \tag{7.21}$$

Thus, Chirikov's parameter $\bar{K}$ for the overlap of the NRs has the value

$$\bar{K} = \frac{\delta J}{\Delta I} = \frac{2\sqrt{2\varepsilon\mu\sqrt{I_0}}}{\nu_0}. \tag{7.22}$$

These estimates are derived in the approximation of isolated NRs ($\bar{K} \ll 1$). In the opposite case the ($\bar{K} > 1$) NRs are interacting strongly, and this leads to the dynamical chaos in region of phase space which is bounded in action.

7.3 The Results of Numerical Simulations of Quantum Dynamics

In the numerical experiment we compute the time-scale $\tau_\hbar$ at which quantum dynamics starts to differ significantly from classical. In addition, we compute the time dependence of quantum correlations in the region of quantum chaos. The time-scale $\tau_\hbar$ is determined in this simulations when the relative diference of the quantum solution from the classical one reaches 2%, i.e.,

$$|\beta_q(\tau_\hbar) - \beta_{cl}(\tau_\hbar)| = 0.02|\beta_{cl}(\tau_\hbar)|, \tag{7.23}$$

where indices cl and q refer to the classical and quantum motions, respectively. In carrying out this numerical experiment with two interacting QNRs the following parameters were chosen

$$\mu = 1, \quad \varepsilon = 0.1, \quad \omega_0 - \omega = 20, \quad I_0 = 10, \quad \nu_0 = 0.5. \tag{7.24}$$

For comparison we considered also the integrable case of isolated QNR which formally can be derived from (7.9) with $\nu_0 = 0$ (by matching of two QNRs; in this case in (7.2) ε should be substituted by 2ε). In the case of the chosen parameters (7.24) we have from (7.22)

$$\bar{K} \approx 3.2, \quad \frac{2\nu_0}{(\omega_0 - \omega)} = \frac{1}{20}, \quad \frac{\delta J}{I_0} \approx 0.16. \tag{7.25}$$

Fig. 3 shows the result of numerical calculations of the time-scale $\tau_\hbar$ for two interacting QNRs according to criterion (7.23). The horizontal axis plots the logarithm of the quasiclassical parameter, which was chosen (as all parameters in (7.24)) in dimensionless form

$$\frac{1}{\hbar} = 2^n, \quad (n = 2, 4, ..., 30). \tag{7.26}$$

In this case the real quasiclassical parameter reaches the value

$$\left(\frac{I_0}{\hbar}\right)_{max} = 10 \cdot 2^{30} \approx 10^{10},$$

and the number of levels trapped into each QNR is of the order

$$\delta n = \frac{\delta J}{\hbar} = \delta J \cdot 2^n,$$

where δJ can be taken from (7.21).

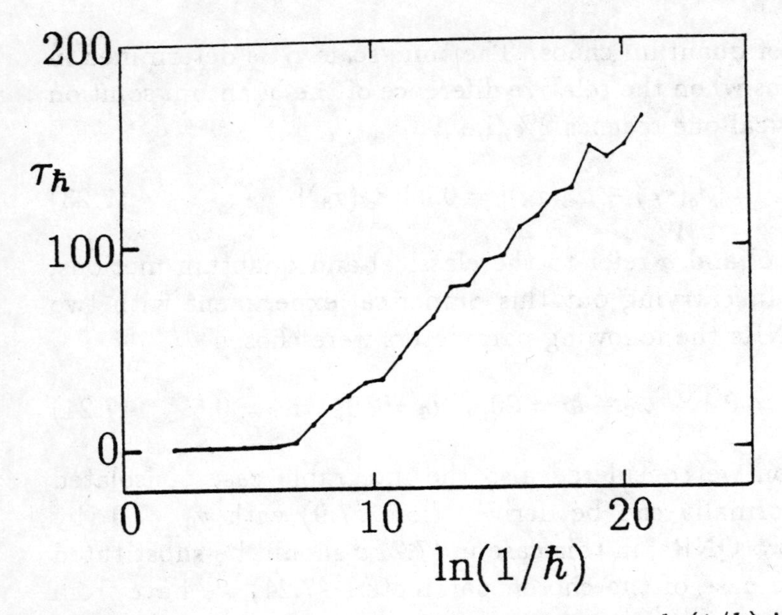

Fig. 3. Dependence of the time-scale $\tau_\hbar$ on $\ln(1/\hbar)$ in the case of chaotic behavior in the classical limit; $\mu = 1; \varepsilon = 0.2; \omega_0 - \omega = 20; I_0 = 10; 1/\hbar = 2^n; (n = 2, ..., 30)$.

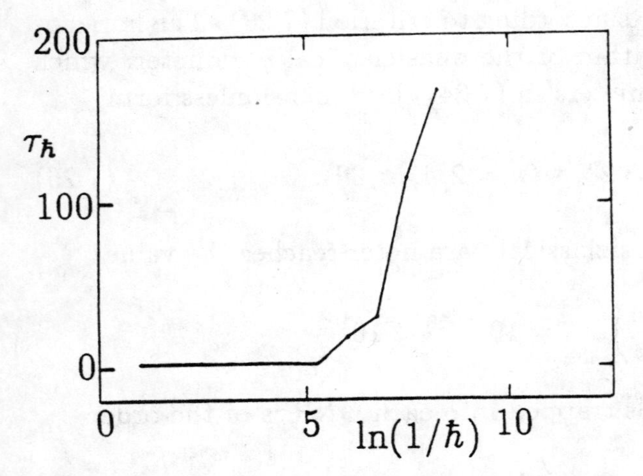

Fig. 4. Dependence $\tau_\hbar$ on $\ln(1/\hbar)$ derived from the criterion (7.23) for the regular motion in the classical limit ($\nu_0 = 0$, isolated QNR). All parameters are the same as in Fig. 3.

Thus, $(\delta n)_{max} \sim 10^9$ levels. The initial conditions in Fig. 3 are the following: the amplitude $|\beta_q(0)| = \sqrt{10}$; and the averaging was made with respect to the initial phase θ_0 ($\beta_q(0) = |\beta_q(0)| \exp(-i\theta_0)$) over twenty trajectories with

$$\theta_0^{(j)} = 2\pi j/20, \quad (j = 0, ..., 19).$$

As seen in Fig. 3, the dependence of $\tau_\hbar$ on the quasiclassical parameter is close to that in (1.2).

For this system of two interacting QNRs it is shown in [9] that the length of the boundary (perimeter) of the initial Wigner-type function grows exponentially during the time $\tau < \tau_\hbar$ (1.2), and for $\tau > \tau_\hbar$ the growth of this length is considerably slowed. Thus, even in the case of the minimum (two) ovelapped primary QNRs the difference between dynamics of quantum expectation values and the classical chaotic dynamics exhibits itself in logarithmically small times. For comparison in Fig. 4 the dependence $\tau_\hbar$ on $\ln(1/\hbar)$ is shown for the integrable case of isolated QNR ($\nu_0 = 0$). As one can see the time-scale $\tau_\hbar$ increases significantly faster with quantum parameter in the integrable case. Finally, we consider the dynamics of the quantum expectation values that determine the purely quantum correlations, and are equal to zero in the classical limit for all time. As an example we choose the following quantum correlator

$$P(t) =< \alpha_0|\mathbf{a}^+(t)\mathbf{a}(t)|\alpha_0 > - < \alpha_0|\mathbf{a}^+(t)|\alpha_0 >< \alpha_0|\mathbf{a}(t)|\alpha_0 > . \tag{7.28}$$

Fig. 5 gives the result of numerical calculation of the dynamics of quantum correlator (7.28) for the chaotic motion in the classical limit. It is seen from Fig. 5, that the quantum correlator $P(t)$ grows in this case in time according to an exponential law. Different dynamical and spectral properties of two interacting QNRs are considered in detail in [15,53,63,64] (see also references in [53]).

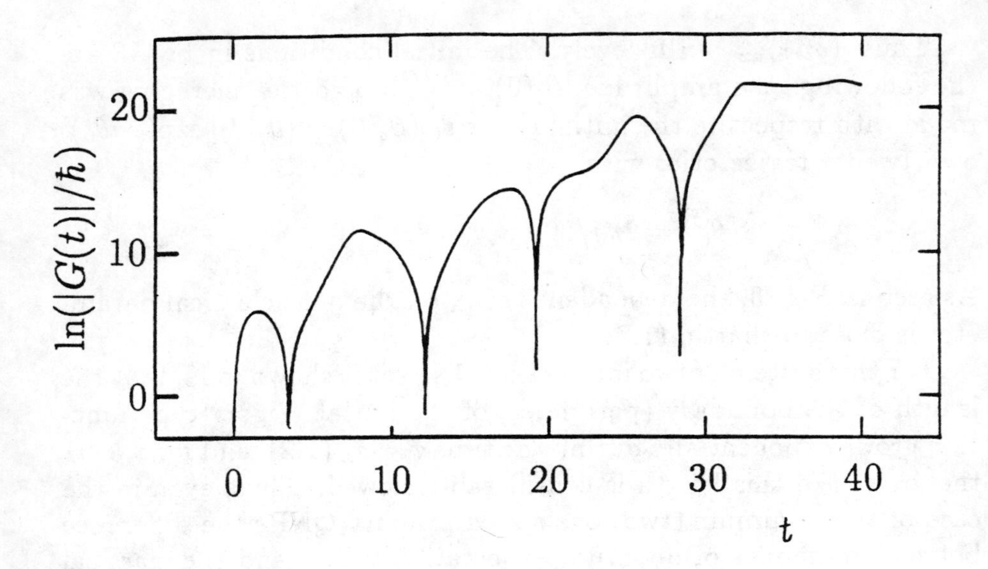

Fig. 5. Time dependence of quantum correlator $G(t)$ (7.28); $\nu_0 = 0.5; \beta_q(0) = -\sqrt{10}$. All other parameters are the same as in Fig. 3.

7.4 Conclusion

The method considered above of calculating quantum dynamics for a system of two interacting QNRs appears to be useful when estimating the rate at which quantum effects grow in stable and chaotic regimes. Under chaotic motion in the classical limit, the growth of quantum effects is exponential in time. The time-scale $\tau_\hbar$ at which quantum dynamics begins to deviate from the classical chaotic motion depends neally logarithmically on the quasiclassical parameter, in agreement with (1.2). The numerical method can also be used for more complicated quantum nonlinear systems, including systems with many degrees of freedom. Since the number of overlapping primary nonlinear resonances in real experiments is usually finite, the verification of the time-scale $\tau_\hbar$ and time-dependence of quantum correlations in experimental systems in the deep quasiclassical region is of significant interest.

8 Two Interacting Quantum Nonlinear Resonances in a Spin System

In this section we consider a quantum spin system of two interacting QNRs exhibiting chaotic dynamics in the classical limit. To analyze its quantum dynamics we use the method discussed in section (3.8). We show that under conditions of developed chaos in the classical limit the quantum correlation functions again grow exponentially in time, and the time-scale $\tau_\hbar$ has the logarithmic dependence (1.2) in this case, as well.

8.1 Introduction

Consider a quantum spin system exhibiting chaotic behavior in its classical limit. As a model, we choose its Hamiltonian in the following form [7]

$$\mathbf{H} = \mathbf{H}_0 - \varepsilon\hbar\mathbf{S}^x\cos(\omega t), \quad \mathbf{H}_0 = D\hbar^2\left(\mathbf{S}^z\right)^2 . \tag{8.1}$$

The Hamiltonian $\mathbf{H}_0$ is considered in section 5.4 (see (5.59)); $\mathbf{S} = (\mathbf{S}^x, \mathbf{S}^y, \mathbf{S}^z)$ are spin operators with the commutators (2.49). The length of the spin is arbitrary (see (5.63)). In (8.1) D is a constant of anisotropy (as in (5.59)); ε and ω are the amplitude and frequency of an external field. Classical consideration corresponds to the limit (5.69). In [7] the peculiarities of quantum dynamics are considered when in the classical limit (5.69) chaotic behaviour occurs as a result of overlapping of NRs. Physically, two primary NRs appear in the system with the Hamiltonian (8.1) because of the interaction of spin rotation with right and left polarizations of the external field. Our approach will be the following. We construct for the system (8.1) a perturbation theory of the type discussed in section 3, and apply it to analyze dynamics in the region of parameters of quantum chaos. The first step consists in deriving a closed set of c-number equations for expectation values in spin CSs. Then by using perturbation theory in quasiclassical region ($S \gg 1$) we construct the solution of these equations and discuss the peculiarities of their quantum dynamics.

8.2 Time-Independent Hamiltonian and Closed C-Number Equations

In section 3.5 a closed c-number equation is derived that describes the dynamics of expectation values for an arbitrary *time - independent* spin Hamiltonian, provided at the initial time spin CSs are chosen. For the case (8.1) it is impossible to apply directly the method discussed in section 3.5, because the Hamiltonian (8.1) depends explicitly on time. So, first we consider instead of (8.1) an effective time-independent Hamiltonian (see section 3.8). For this we introduce instead of the time-periodic external field in (8.1) an auxiliaries boson field described by boson operators $\mathbf{b}$ and $\mathbf{b}^+$. Thus, instead of (8.1) we consider first the following Hamiltonian

$$\mathbf{H}_{eff} = \hbar\omega\mathbf{b}^+\mathbf{b} + D\hbar^2\left(\mathbf{S}^z\right)^2 - \frac{\hbar\varepsilon}{2\beta_0}\left(\mathbf{b}^+ + \mathbf{b}\right)\mathbf{S}^x, \quad \left([\mathbf{b},\mathbf{b}^+] = 1\right), \tag{8.2}$$

where β_0 is a real constant. Assume that at $t = 0$ the field described by the operator $\mathbf{b}$ is in CS $|\beta_0>$, and the energy of this field is large enough that

$$\hbar\omega|\beta_0|^2 \gg \left\{DS_0^2, \varepsilon S_0\right\}, \quad (S_0 = \hbar S). \tag{8.3}$$

Then, according to (8.2) we have in the first approximation

$$\mathbf{b}(t) \approx e^{-i\omega t}\mathbf{b}(0). \tag{8.4}$$

As $\mathbf{b}(0)|\beta_0>= \beta_0|\beta_0>$, we derive from (8.2) under the condition (8.3) the Hamiltonian (8.1). We shall come to this problem again in section 12.3. For the time-independent Hamiltonian (8.2) we can apply a method based on closed c-number equations for the expectation values in spin and boson CSs. We introduce for this spin CS $|\xi>$ in (2.51). Since Hamiltonian (8.2) includes also the boson operators $\mathbf{b}^+$ and $\mathbf{b}$, we introduce at $t = 0$ the boson CS,

$$|\beta>= e^{\beta\mathbf{b}^+ - \beta^*\mathbf{b}}|0>= e^{-|\beta|^2/2}\sum_{n=0}^{\infty}\frac{\beta^n}{\sqrt{n!}}|n>, \tag{8.5}$$

$$\left(\mathbf{b}^+\mathbf{b}|n>= n|n>, \quad \mathbf{b}|\beta>= \beta|\beta>\right).$$

Next, we derive a closed c-number equation for the expectation value

$$f(t) =< \xi,\beta|\mathbf{f}(t)|\beta,\xi >= f(\xi^*,\xi;\beta^*,\beta,t) \tag{8.6}$$

in the state

$$|\beta,\xi >\equiv |\beta > |\xi >$$

of an arbitrary Heisenberg operator

$$\mathbf{f}(t) \equiv \mathbf{f}\left[\mathbf{S}^z(t),\mathbf{S}^+(t),\mathbf{S}^-(t),\mathbf{b}^+(t),\mathbf{b}(t)\right]. \tag{8.7}$$

By using the method discussed in section 3, we have for the expectation value $f(t)$ (8.6) the linear partial differential equation

$$\dot{f} = \hat{K} f, \quad f(0) = f(\xi^*,\xi;\beta^*,\beta), \tag{8.8}$$

where the differential operator $\hat{K}$ in (8.8) is given by the expression

$$\hat{K} = \frac{i}{\hbar} e^{-|\beta|^2}\left(1+|\xi|^2\right)^{-2S} \{\mathbf{H}_{eff}(\mathbf{S}^z = \xi^*\frac{\partial}{\partial\xi^*} - S, \mathbf{S}^+ = 2S\xi^* - \xi^{*2}\frac{\partial}{\partial\xi^*}, \tag{8.9}$$

$$\mathbf{S}^- = \frac{\partial}{\partial\xi^*}, \mathbf{b}^+ = \beta^*, \mathbf{b} = \frac{\partial}{\partial\beta^*}) - \mathbf{H}_{eff}(\mathbf{S}^z = \xi\frac{\partial}{\partial\xi} - S,$$

$$\mathbf{S}^+ = \frac{\partial}{\partial\xi}, \mathbf{S}^- = 2S\xi - \xi^2\frac{\partial}{\partial\xi}, \mathbf{b}^+ = \frac{\partial}{\partial\beta}, \mathbf{b} = \beta)\}$$

$$\times (1+|\xi|)^{2S} e^{|\beta|^2}.$$

For the Hamiltonian $\mathbf{H}_{eff}$ (8.2) the operator $\hat{K}$ in (8.9) takes the explicit form

$$\hat{K} = \frac{i}{\hbar} e^{-|\beta|^2}\left(1+|\xi|^2\right)^{-2S} \tag{8.10}$$

$$\times\{\hbar\omega\left(\beta^*\frac{\partial}{\partial\beta^*} - \beta\frac{\partial}{\partial\beta}\right) + D\hbar^2\left(\xi^*\frac{\partial}{\partial\xi^*} - S\right)^2 - D\hbar^2\left(\xi\frac{\partial}{\partial\xi} - S\right)^2$$

$$-\frac{\hbar\varepsilon}{4\beta_0}\left(\beta^* + \frac{\partial}{\partial\beta^*}\right)\left(2S\xi^* - \xi^{*2}\frac{\partial}{\partial\xi^*} + \frac{\partial}{\partial\xi^*}\right)$$

$$+\frac{\hbar\varepsilon}{4\beta_0}\left(\beta + \frac{\partial}{\partial\beta}\right)\left(2S\xi - \xi^2\frac{\partial}{\partial\xi} + \frac{\partial}{\partial\xi}\right)\}\left(1+|\xi|^2\right)^{2S} e^{|\beta|^2}$$

$$
\begin{aligned}
&= i\{\omega\left(\beta^*\frac{\partial}{\partial\beta^*} - \beta\frac{\partial}{\partial\beta}\right) - \frac{S\varepsilon(\xi^*+\xi)}{2\beta_0(1+|\xi|^2)}\left(\frac{\partial}{\partial\beta^*} - \frac{\partial}{\partial\beta}\right) \\
&\quad - \frac{\varepsilon(\beta^*+\beta)}{4\beta_0}\left[(1-\xi^{*2})\frac{\partial}{\partial\xi^*} - (1-\xi^2)\frac{\partial}{\partial\xi}\right] \\
&\quad - \frac{\varepsilon}{4\beta_0}\left[(1-\xi^{*2})\frac{\partial^2}{\partial\xi^*\partial\beta^*} - (1-\xi^2)\frac{\partial^2}{\partial\xi\partial\beta}\right] \\
&\quad + D\left[\frac{4S_0|\xi|^2}{(1+|\xi|^2)} - \hbar(2S-1)\right]\left(\xi^*\frac{\partial}{\partial\xi^*} - \xi\frac{\partial}{\partial\xi}\right) \\
&\quad + D\hbar\left(\xi^{*2}\frac{\partial^2}{\partial\xi^{*2}} - \xi^2\frac{\partial^2}{\partial\xi^2}\right)\}.
\end{aligned}
$$

In the next section we shall simplify the operator $\hat{K}$ in (8.10) by using the condition of "constant amplitude of the external field" (3.65).

8.3 The Approximation of Constant Amplitude of the External Field

In the limit (3.65) we pass from (8.2) over to the initial Hamiltonian (8.1). We show how this limit works for equations (8.8)-(8.10). In the limit $\beta_0 \to \infty$ we neglect in (8.10) the terms involving derivatives $\partial/\partial\beta$, $\partial/\partial\beta^*$ in comparison with the terms $\beta(\partial/\partial\beta)$, $\beta^*(\partial/\partial\beta^*)$. In this approximation we have from (8.10) that

$$
\hat{K} = \hat{K}_\beta + \hat{K}_\xi, \tag{8.11}
$$

where

$$
\begin{aligned}
\hat{K}_\beta &= i\{\omega\left(\beta^*\frac{\partial}{\partial\beta^*} - \beta\frac{\partial}{\partial\beta}\right) \\
&\quad - \frac{\varepsilon(\beta^*+\beta)}{4\beta_0}\left[(1-\xi^{*2})\frac{\partial}{\partial\xi^*} - (1-\xi^2)\frac{\partial}{\partial\xi}\right],
\end{aligned} \tag{8.12}
$$

and

$$
\hat{K}_\xi = iD\left[\frac{4S_0|\xi|^2}{(1+|\xi|^2)} - \hbar(2S-1)\right]\left(\xi^*\frac{\partial}{\partial\xi^*} - \xi\frac{\partial}{\partial\xi}\right) \tag{8.13}
$$

$$+iD\hbar\left(\xi^{*2}\frac{\partial^2}{\partial\xi^{*2}}-\xi^2\frac{\partial^2}{\partial\xi^2}\right).$$

The function $f(t)$ in (8.8) depends on the spin and boson variables (see (8.6)): $\xi^*, \xi; \beta^*, \beta$. We transform variables β and β^* to a rotating system of coordinates with frequency ω. For this, we use the substitution

$$\tilde{\beta}^* = e^{i\omega t}\beta^*, \quad \tilde{\beta} = e^{-i\omega t}\beta. \tag{8.14}$$

In the new variables we have instead of the operator $\hat{K}_\beta$ (8.12)

$$\hat{K}_{\tilde{\beta}} = -i\frac{\varepsilon\left(e^{-i\omega t}\tilde{\beta}^* + e^{i\omega t}\tilde{\beta}\right)}{4\beta_0}\left[(1-\xi^{*2})\frac{\partial}{\partial\xi^*}-(1-\xi^2)\frac{\partial}{\partial\xi}\right]. \tag{8.15}$$

Because the operator $\hat{K}_{\tilde{\beta}}$ does not include the derivatives by $\tilde{\beta}$ and $\tilde{\beta}^*$, one may consider $\tilde{\beta}$ and $\tilde{\beta}^*$ as parameters, and choose them as follows: $\tilde{\beta} = \tilde{\beta}^* = \beta_0$. Thus, we have from (8.15)

$$\hat{K}_{\beta_0} = -i\frac{\varepsilon\cos(\omega t)}{2}\left[(1-\xi^{*2})\frac{\partial}{\partial\xi^*}-(1-\xi^2)\frac{\partial}{\partial\xi}\right]. \tag{8.16}$$

Upon using (8.13) and (8.16), equation (8.8) becomes

$$\dot{f} = \hat{K}f, \quad \hat{K} = \hat{K}_{cl} + \frac{1}{S}\hat{K}_q, \tag{8.17}$$

where $\hat{K}_{cl}$ includes only first order derivatives and describes the classical limit (5.69)

$$\hat{K}_{cl} = iD\left[\frac{4S_0|\xi|^2}{(1+|\xi|^2)} - \hbar(2S-1)\right]\left(\xi^*\frac{\partial}{\partial\xi^*}-\xi\frac{\partial}{\partial\xi}\right) \tag{8.18}$$

$$-i\frac{\varepsilon\cos(\omega t)}{2}\left[(1-\xi^{*2})\frac{\partial}{\partial\xi^*}-(1-\xi^2)\frac{\partial}{\partial\xi}\right].$$

The operator $\hat{K}_q$ describes the quantum effects

$$\hat{K}_q = iDS_0\left(\xi^{*2}\frac{\partial^2}{\partial\xi^{*2}}-\xi^2\frac{\partial^2}{\partial\xi^2}\right). \tag{8.19}$$

In (8.18) we included the term of order $\sim\hbar$ in the classical operator, since it involves only first order derivatives.

8.4 Time-Dependent Quantum Correlation Functions

Following section 3.7 we consider two arbitrary operators **f** and **g**, and introduce for them the quantum correlation function (QCF)

$$\frac{P(t)}{S} \equiv <\xi,\beta|\mathbf{f}(t)\mathbf{g}(t)|\beta,\xi> - <\xi,\beta|\mathbf{f}(t)|\beta,\xi><\xi,\beta|\mathbf{g}(t)|\beta,\xi>. \tag{8.20}$$

As seen from definition (8.20) the function $P(t)$ describes the purely quantum correlations. In this section we consider a perturbation theory for describing the dynamics of the QCF $P(t)$ in various regions of parameters. First we write an exact equation for $P(t)$ using the following equations for $f(t)$, $g(t)$ and $<fg>$ (see (3.59))

$$\frac{\partial <fg>}{\partial t} = \hat{K} <fg>, \quad \frac{\partial f}{\partial t} = \hat{K} f, \quad \frac{\partial g}{\partial t} = \hat{K} g, \tag{8.21}$$

with the operator $\hat{K}$ defined in (8.17). Using (8.21) and the explicit form of the operator $\hat{K}_{cl}$ (8.18) and $\hat{K}_q$ (8.19) we have for QCF $P(t)$ (8.20) the following equation

$$\frac{\partial P(t)}{\partial t} = \hat{K}_{cl}P(t) + 2iDS_0\left(\xi^{*2}\frac{\partial f}{\partial \xi^*}\frac{\partial g}{\partial \xi^*} - \xi^2\frac{\partial f}{\partial \xi}\frac{\partial g}{\partial \xi}\right) + \frac{1}{S}\hat{K}_q P(t). \tag{8.22}$$

Now we construct an approximate solution for $P(t)$ using the perturbation expansion in the small parameter $1/S \ll 1$, suggested in section 3.7. For this we neglect in (8.22) the last term, and substitute for functions $f(\xi^*,\xi,t)$ and $g(\xi^*,\xi,t)$ the corresponding classical functions satisfying

$$\frac{\partial f_{cl}(t)}{\partial t} = \hat{K}_{cl} f_{cl}(t), \quad \frac{\partial g_{cl}(t)}{\partial t} = \hat{K}_{cl} g_{cl}(t). \tag{8.23}$$

Thus, we have an approximate equation for the QCF $P(t)$ given by

$$\frac{\partial P}{\partial t} = \hat{K}_{cl}P + D(\xi^*,\xi,t), \quad P \equiv P(\xi^*,\xi,t), \tag{8.24}$$

where the function $D(\xi^*, \xi, t)$ in (8.24) is considered to be known, and according to (8.22) has the form

$$D(\xi^*, \xi, t) = 2iDS_0 \left(\xi^{*2} \frac{\partial f_{cl}}{\partial \xi^*} \frac{\partial g_{cl}}{\partial \xi^*} - \xi^2 \frac{\partial f_{cl}}{\partial \xi} \frac{\partial g_{cl}}{\partial \xi} \right). \tag{8.25}$$

Since the operator K_{cl} includes only the first order derivatives, and the function D in (8.24) is supposed to be known, the solution of equation (8.24) can easily be found by the method of characteristics (see section 4.1).

8.5 Quantum Perturbation Theory

Choose some concrete operators $\mathbf{f}$ and $\mathbf{g}$. Let, for example,

$$\mathbf{f} \equiv \mathbf{J}^+, \quad \mathbf{g} \equiv \mathbf{J}^z, \tag{8.26}$$

where

$$\mathbf{J}^+ = \mathbf{S}^+/S, \quad \mathbf{J}^- = \mathbf{S}^-/S, \quad \mathbf{J}^z = \mathbf{S}^z/S. \tag{8.27}$$

According to (8.20) we introduce the QCF

$$\frac{\Lambda^{+,z}(t)}{S} = < \xi|\mathbf{J}^+(t)\mathbf{J}^z(t)|\xi > - < \xi|\mathbf{J}^+(t)|\xi >< \xi|\mathbf{J}^z(t)|\xi >, \tag{8.28}$$

where instead of the general definition for QCF $P(t)$ used in (8.20), we use the definition (8.28) for this QCF. At the initial time $t = 0$ we have

$$\Lambda^{+,z}(t = 0) \equiv \Lambda_0^{+,z} = -\frac{2|\xi|^2\xi^*}{(1 + |\xi|^2)^2}. \tag{8.29}$$

As was mentioned above, we have in the classical limit for QCF (8.28)

$$\lim_{\hbar \to 0} \frac{\Lambda^{+,z}(t)}{S} = \lim_{\hbar \to 0} \frac{\hbar}{S_0} \Lambda^{+,z}(t) = 0. \tag{8.30}$$

Thus, this QCF describes purely quantum effects. According to (8.24) and (8.25) we present an equation for $\Lambda^{+,z}(t)$ in the form

$$\frac{\partial \Lambda^{+,z}(t)}{\partial t} = \hat{K}_{cl} \Lambda^{+,z}(t) + 2iDS_0 \tag{8.31}$$

$$\left(\xi^{*2}\frac{\partial J^+_{cl}(t)}{\partial\xi^*}\frac{\partial J^z_{cl}(t)}{\partial\xi^*}-\xi^2\frac{\partial J^+_{cl}(t)}{\partial\xi}\frac{\partial J^z_{cl}(t)}{\partial\xi}\right).$$

As the last term in (8.31) is known, integration of equation (8.31) can be done by the method of characteristics, as discussed in section 4.1. First consider the equation

$$\frac{\partial\Lambda^{+,z}(t)}{\partial t}=\hat{K}_{cl}\Lambda^{+,z}(t). \tag{8.32}$$

According to (8.18) we have from (8.32) equation for characteristics

$$\dot{\varphi}(\vec{x},t)=\frac{i\varepsilon}{2}\Big(1-\varphi^2(\vec{x},t)\Big)\cos(\omega t) \tag{8.33}$$

$$-iD\left[\frac{4S_0|\varphi(\vec{x},t)|^2}{(1+|\varphi(\vec{x},t)|^2)}-\hbar(2S-1)\right]\varphi(\vec{x},t),$$

where

$$\vec{x}=(x_1,x_2)\equiv(\xi,\xi^*).$$

The solution of (8.31) presents itself in the form

$$\Lambda^{+,z}(\vec{x},t)=\Lambda^{+,z}\left(\varphi(\vec{x},t),\varphi^*(\vec{x},t)\right)+\sum_{\alpha,\beta=1,2}\frac{\partial J^+_{cl}(\vec{x},t)}{\partial x_\alpha}\frac{\partial J^z_{cl}(\vec{x},t)}{\partial x_\beta} \tag{8.34}$$

$$\times[\int_0^t d\tau\{\varphi^{*2}(\vec{x},\tau)T^{-1}_{\alpha,2}(\vec{x},\tau))T^{-1}_{\beta,2}(\vec{x},\tau)$$

$$-\varphi^2(\vec{x},\tau)T^{-1}_{\alpha,1}(\vec{x},\tau))T^{-1}_{\beta,1}(\vec{x},\tau)\}],$$

where

$$T_{i,j}=\begin{pmatrix}\frac{\partial\varphi(\vec{x},\tau)}{\partial x_1}, & \frac{\partial\varphi(\vec{x},\tau)}{\partial x_2}\\ \frac{\partial\varphi^*(\vec{x},\tau)}{\partial x_1}, & \frac{\partial\varphi^*(\vec{x},\tau)}{\partial x_2}\end{pmatrix}, \tag{8.35}$$

and

$$J^+_{cl}(\vec{x},t)=\frac{2\varphi^*(\vec{x},t)}{(1+|\varphi(\vec{x},t)|^2)}, \tag{8.36}$$

$$J^z_{cl}(\vec{x},t)=-\frac{(1-|\varphi(\vec{x},t)|^2)}{(1+|\varphi(\vec{x},t)|^2)}.$$

For the inverse matrix in (8.34) we have from (8.35)

$$T_{i,j}^{-1} = \frac{1}{(|\frac{\partial\varphi(\vec{x},\tau)}{\partial x_1}|^2 - |\frac{\partial\varphi(\vec{x},\tau)}{\partial x_2}|^2)} \tag{8.37}$$

$$\begin{pmatrix} \dfrac{\partial\varphi^*(\vec{x},\tau)}{\partial x_2} & , & -\dfrac{\partial\varphi(\vec{x},\tau)}{\partial x_2} \\ -\dfrac{\partial\varphi^*(\vec{x},\tau)}{\partial x_1} & , & \dfrac{\partial\varphi(\vec{x},\tau)}{\partial x_1} \end{pmatrix},$$

Finally, to calculate QCF (8.34) one needs to know the derivatives

$$\partial\varphi(\vec{x},t)/\partial x_i, \quad (i = 1,2),$$

at an arbitrary time t. These derivatives can be found by solving simultaneously the equations for characteristics $\varphi(\vec{x},t)$ and $\varphi^*(\vec{x},t)$ (8.33) and the system of equations

$$\dot{\varphi}_{,\alpha} = -i\varepsilon\varphi_{,\alpha}\varphi\cos(\omega t) - iD\left[\frac{4S_0|\varphi|^2}{(1+|\varphi|^2)} - \hbar(2S-1)\right]\varphi_{,\alpha} \tag{8.38}$$

$$-4iS_0D\frac{\varphi(\varphi^*\varphi_{,\alpha} + \varphi\varphi^*_{,\alpha})}{(1+|\varphi|^2)^2},$$

where

$$\varphi_{,\alpha} \equiv \partial\varphi(\vec{x},t)/\partial x_\alpha, \quad (\alpha = 1,2).$$

The system of equations (8.38) is derived by differentiation of equation (8.33) by x_α. The initial conditions for the equations (8.33) and (8.38) are the following

$$\varphi(t=0) = \xi, \quad \varphi^*(t=0) = \xi^*, \tag{8.39}$$

$$\frac{\partial\varphi}{\partial x_1}|_{t=0} = 1, \quad \frac{\partial\varphi}{\partial x_2}|_{t=0} = 0, \quad \frac{\partial\varphi^*}{\partial x_1}|_{t=0} = 0, \quad \frac{\partial\varphi^*}{\partial x_2}|_{t=0} = 1.$$

We also present equations which describe the quantum dynamics of expectation values

$$J^+(t) \equiv <\xi|\mathbf{J}^+(t)|\xi>, \quad J^z(t) \equiv <\xi|\mathbf{J}^z(t)|\xi>. \tag{8.40}$$

For this we first write the Heisenberg equations for operators $\mathbf{J}^+(t)$ and $\mathbf{J}^z(t)$. From (8.1) and the definition (8.27) we have the operator equations

$$\dot{\mathbf{J}}^z = i\frac{\varepsilon}{2}\cos(\omega t)\left(\mathbf{J}^+ - \mathbf{J}^-\right), \tag{8.41}$$

$$\dot{\mathbf{J}}^+ = iDS_0\left(2\mathbf{J}^+\mathbf{J}^z + \mathbf{J}^+/S\right) + i\varepsilon\mathbf{J}^z\cos(\omega t).$$

Averaging (8.41) in CS $|\xi>$ we derive the exact c-number equations for expectation values

$$\dot{J}^z = i\frac{\varepsilon}{2}\cos(\omega t)\left(J^+ - J^-\right), \tag{8.42}$$

$$\dot{J}^+ = i\varepsilon J^z\cos(\omega t) + iDS_0\left(J^+/S + 2\Lambda^{+,z}/S + 2J^+J^z\right),$$

where the QCF $\Lambda^{+,z}(t)$ is defined in (8.28). For the approximate solution of (8.42) we can instead of the exact QCF $\Lambda^{+,z}(t)$ use its approximation given by (8.34). In this case we derive from (8.42) equations including quantum terms of order $1/S$. Further, we call these solutions quasiclassical: $J^z_{qc}(t)$; $J^+_{qc}(t)$. To derive the approximate solutions $J^z_{qc}(t)$ and $J^+_{qc}(t)$ one can use also as an initial step the quantum equation (8.17). For example,

$$\frac{\partial J^+}{\partial t} = \hat{K}_{cl}J^+ + \frac{1}{S}\hat{K}_q J^+, \tag{8.43}$$

$$J^+(0) = \frac{2\xi^*}{1+|\xi|^2},$$

where $\hat{K}_{cl}$ and $\hat{K}_q$ are given in (8.18) and (8.19). We present the solution $J^+(t)$ in the form

$$J^+ = J^+_{cl} + \frac{1}{S}J^+_q, \tag{8.44}$$

where

$$\dot{J}_{cl} = \hat{K}_{cl}J^+_{cl}. \tag{8.45}$$

Then, we have for J^+_q the exact equation

$$\frac{\partial J^+_q}{\partial t} = \hat{K}_{cl}J^+_q + \hat{K}_q J^+_{cl} + \frac{1}{S^2}\hat{K}_q J^+_q. \tag{8.46}$$

As we are interested in quantum corrections of order $1/S$ to the classical solution J_{cl}^{+}, we neglect in (8.46) the last term. Then, we have an equation for $J_{qc}^{+}(t)$

$$\frac{\partial J_{qc}^{+}}{\partial t} = \hat{K}_{cl} J_{qc}^{+} + \hat{K}_{q} J_{cl}^{+}, \tag{8.47}$$

where the last term is considered to be known (from the equation (8.45)). The operator $\hat{K}_{cl}$ in (8.47) again includes only first order derivatives, and so may be solved by the method of characteristics.

The integrable case $\varepsilon = 0$ for the Hamiltonian (8.1) is considered in section 5.4. When $\varepsilon \neq 0$ there exists a region of parameters (see [7]), where global chaos occurs in the classical limit due to overlapping NRs. In this case in [7] the time-scale $\tau_{\hbar}$ was numerically calculated and is shown to be of the order

$$\tau_{\hbar} \approx \frac{1}{\lambda_L} \ln S, \quad (S = \frac{S_0}{\hbar}), \tag{8.48}$$

where λ_L is a characteristic Lyapunov exponent. For numerical calculations it is more convenient to use the equations (8.34) and (8.42), as in this case only the first order derivatives in (8.34) need to be numerically calculated. To calculate numerically the dependence $\tau_{\hbar}(S)$ we used the criterion

$$\delta(\tau_{\hbar}) = \left(|J_{qc}^{+}(t) - J_{cl}^{+}(t)|^2 + |J_{qc}^{z}(t) - J_{cl}^{z}(t)|^2\right)|_{t=\tau_{\hbar}} = constant \ll 1, \tag{8.49}$$

where a *constant* in (8.49) is choosen in numerical calculations: *constant* = *0.02* ("2% criterion"). We take $\delta(0) = 0$. Then, the "distance" between classical and quasiclassical solutions at $t = \tau_{\hbar}$ increases up to $\delta(\tau_{\hbar}) = 0.02$. Fig. 6 shows the dependence $\tau_{\hbar}$ on $\ln S$ under condition of chaotic motion in the classical limit (which is illustrated in Fig. 7).

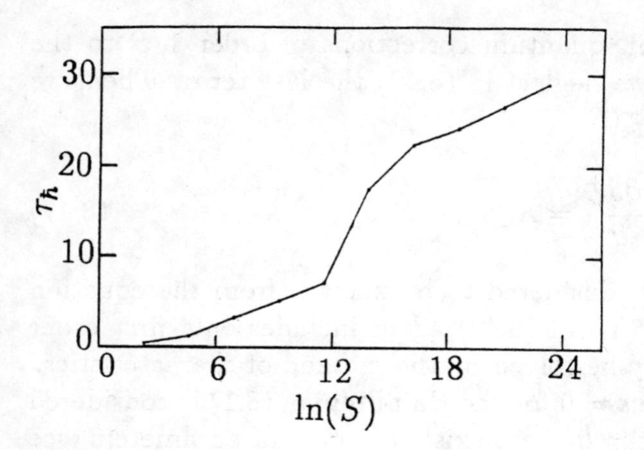

Fig. 6. Dependence $\tau_\hbar$ on $\ln S$ for chaotic motion in the classical limit; $\varepsilon = 0.3; \omega = 1; D = 1.4; S_0 = 1; \xi = 0.7(1+i)$.

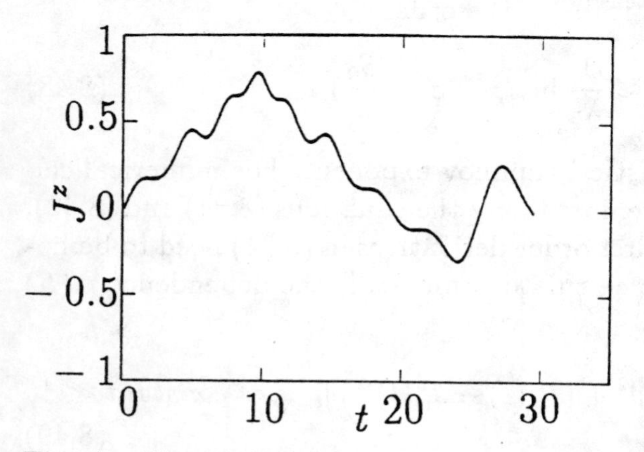

Fig. 7. Classical chaotic motion ($\hbar = 0; S = \infty; \hbar S = S_0 = 1$). All other parameters are the same as in Fig. 6.

As it is seen from Fig. 6 the dependence $\tau_\hbar$ on S varies according to (8.48). For comparison in Fig. 8 the dependence $\tau_\hbar$ on S is given for the regular motion ($\varepsilon = 0$).

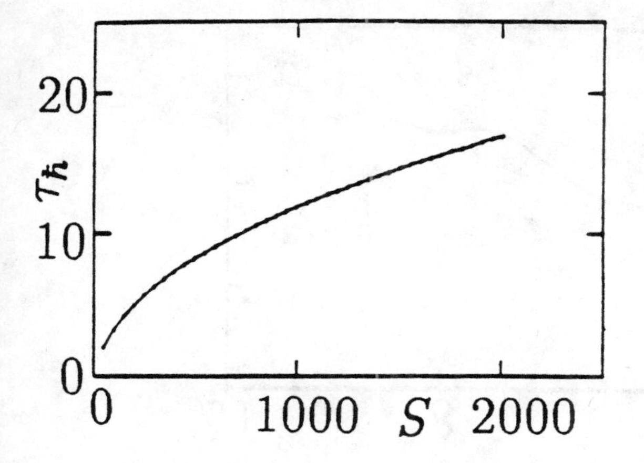

Fig. 8. Dependence $\tau_\hbar$ on S for the regular classical motion; $\varepsilon = 0; S_0 = 1; D = -1; \xi = 3 + 2i$.

In this case the dependence $\tau_\hbar$ on S is approximately an algebraic power law. The time behaviour of QCF $\Lambda^{+,z}(t)$ (8.28) is shown in Figs. 9a,b. The dependence shown in Fig. 9a corresponds to the chaotic classical motion. As one can see quantum correlations grow exponentially in time in this case. For regular classical motion quantum correlations grow powerwise (Fig. 9b).

In conclusion we note, that for the system with Hamiltonian (8.1) there exists a significant difference in the behaviour of QCFs in the quasiclassical region ($S \gg 1$) for regular and chaotic motion in the corresponding classical limit. Namely, under conditions of developed chaos in the classical limit, QCFs grow exponentially in time, and the time-scale $\tau_\hbar$ is logarithmically small by the quasiclassical parameter κ, which in this case coincides with the dimensionless spin length S (see (5.77)). Under conditions of regular classical dynamics the QCFs grow powerwise, and the time-scale $\tau_\hbar$ has a functional dependence on the quasiclassical parameter close to (1.1).

There is significant interest in a "classical-quantum" transitional region of dynamical behaviour of the systems of the type (8.1) when $\tau \sim \tau_\hbar$. Investigations of this transition for nonintegrable Hamiltonian spin systems with 1.5 degrees of freedom in quasiclassical region of parameters are carried out in [8,11].

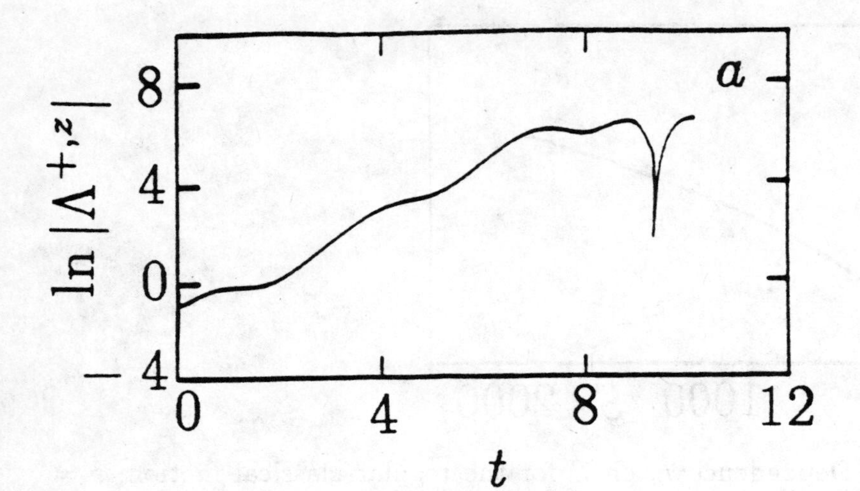

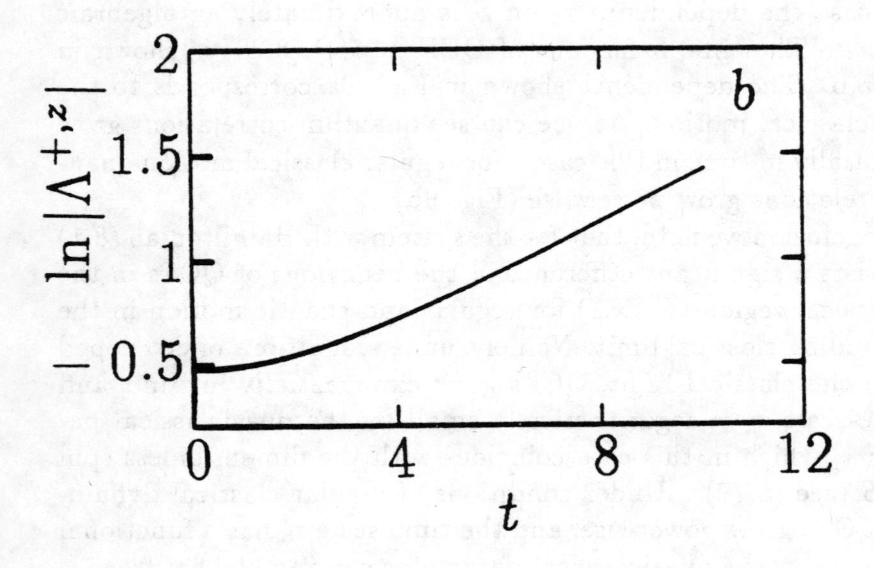

Fig. 9. Dependence of quantum correlator $\Lambda^{+,z}(t)|$ on time for developed chaotic (a), and for regular motion (b) in the classical limit: (a) $\varepsilon = 0.3; S_0 = 1; \omega = 1; \xi = 0.7(1+i); D = 1.4; S = 10^4$; (b) $\varepsilon = 0; S_0 = 1; \omega = 1; \xi = 3 + 2i; D = -1.4; S = 10^3$.

9 Quantization of a Stochastic Web

9.1 Introduction

In this section we consider following papers [21,68] a quantum system consisting of a quantum linear oscillator kicked by a periodic sequence of δ - pulses. We write the Hamiltonian as

$$\mathbf{H} = -\frac{\hbar^2}{2m_0}\frac{\partial^2}{\partial x^2} + \frac{m_0\omega_0^2 x^2}{2} + \varepsilon\cos(k_0 x)\sum_{n=-\infty}^{\infty}\delta(t - nT_0). \quad (9.1)$$

The difference in system (9.1) from the systems considered above is that the unperturbed system (9.1) ($\varepsilon = 0$) is linear. At resonance

$$T \equiv \omega_0 T_0 = \frac{2\pi r}{q} \quad (9.2)$$

(r,q are integers) the effect of perturbation is especially strong. Recent investigations of system (9.1) in the classical limit ($\hbar = 0$) have led to the discovery of an unbounded stochastic web with symmetry q [55,56]. If one ignores the trivial cases $q = 1, 2$, at $q = 3, 4, 6$ the web will have a periodic structure with the symmetry corresponding to a crystalline lattice. For all other values of q it will have a quasicrystalline structure. These unusual properties of classical chaos make the system (9.2) a new and interesting object for investigating quantum chaos, since the existence of the web affects quantum diffusion of particles.

In this section we mainly compare classical and quantum dynamics of the system (9.1) in the region of parameters for quantum chaos. In section 9.2 we construct a discrete quantum map for the wavefunctions of the system. Section 9.3 briefly describes the properties of the system (9.1) in the classical limit; here also some parameters are introduced which are needed for further analysis. In section 9.4 we derive a recursion equation for a generating function, which allows one to calculate arbitrary time-dependent expectation values.

Section 9.5 deals with the method of determining parameters of the system (9.1) for which at finite times quantum dynamics remains similar to classical dynamics. In section 9.6 the time-scale $\tau_\hbar$ of the quasiclassical description is determined under the condition of strong chaos in the classical limit. This time-scale is shown to be logarithmically small as a function of the quasiclassical parameter. Section 9.7 gives a comparative analysis of the classical and quantum dynamics under the condition of weak chaos in the classical limit. In section 9.8 we present some additional remarks concerning the main results and discuss the possibility of further investigation of the model (9.1).

9.2 Quantum Map

To investigate the properties of system (9.1) it is convenient to pass from the Schrödinger equation to the finite difference equation for a wavefunction $\psi(t)$ by relating its value between two successive δ-pulses, e.g. the n-th and the $(n+1)$-th. This relation may be expressed in the form

$$\psi_{n+1} = \hat{T}\psi_n, \tag{9.3}$$

where

$$\psi_n \equiv \psi(t_n - 0), \quad \psi_{n+1} \equiv \psi(t_{n+1} - 0), \quad t_n \equiv nT_0,$$

and the operator $\hat{T}$ defines this map.

For convenience we introduce the following dimensionless variables

$$\kappa = \frac{\varepsilon}{\hbar}, \quad k = k_0 d, \quad d = \sqrt{\frac{\hbar}{m_0\omega_0}}, \quad \tau = \omega_0 t. \tag{9.4}$$

In these variables the Hamiltonian (9.1) takes the following dimensionless form

$$\hat{\mathcal{H}} = \frac{\mathbf{H}}{\hbar\omega_0} = -\frac{1}{2}\frac{\partial^2}{\partial\eta^2} + \frac{1}{2}\eta^2 + \kappa\cos(k\eta)\sum_{n=-\infty}^{\infty}\delta(\tau - nT). \tag{9.5}$$

Now we introduce operators of creation $\mathbf{a}^+$ and annihilation $\mathbf{a}$ ($[\mathbf{a}, \mathbf{a}^+] = 1$) according to the usual formulae

$$\eta = \frac{1}{\sqrt{2}}(\mathbf{a}^+ + \mathbf{a}), \quad p == \frac{i}{\sqrt{2}}(\mathbf{a}^+ - \mathbf{a}) = -i\frac{\partial}{\partial \eta}. \tag{9.6}$$

Then expression (9.5) takes the form

$$\hat{\mathcal{H}} = \mathbf{a}^+\mathbf{a} + \frac{1}{2}\cos\left[(k/\sqrt{2})(\mathbf{a}^+ + \mathbf{a})\right] \sum_{n=-\infty}^{\infty} \delta(\tau - nT). \tag{9.7}$$

Using (9.7) easily produces the evolution operator $\hat{T}$ introduced in (9.3) from the expression

$$\hat{T} = \exp\left\{-i\int_{\tau_n+0}^{\tau_{n+1}-0} \hat{\mathcal{H}} d\tau\right\} \exp\left\{-i\int_{\tau_n-0}^{\tau_n+0} \hat{\mathcal{H}} d\tau\right\} \equiv \hat{T}_0 \hat{T}_{int}, \tag{9.8}$$

$$\hat{T}_0 = \exp\{-iT(\mathbf{a}^+\mathbf{a} + \frac{1}{2})\}; \quad \hat{T}_{int} = \exp\{-i\kappa\cos[(k/\sqrt{2})(\mathbf{a}^+ + \mathbf{a})]\}.$$

The operator $\hat{T}$ defines a quantum map. From it, the wavefunction ψ_n at time $(\tau_n - 0)$ is found from the equation

$$\psi_n = \hat{T}\psi_0, \tag{9.9}$$

where ψ_0 is the initial condition. The operator $\hat{T}$ can be also used to determine the properties of the quasi-energy eigenfunctions.

9.3 Classical Limit

In what follows we need information on the properties of the system (9.7) or (9.1) in the classical limit $\hbar = 0$. In this case we replace operators $\mathbf{a}^+$ and $\mathbf{a}$ by their corresponding c-numbers

$$\mathbf{a} \to \alpha = \sqrt{I/\hbar}\exp(-i\vartheta), \quad \mathbf{a}^+ \to \alpha^* = \sqrt{I/\hbar}\exp(i\vartheta), \tag{9.10}$$

where (I, ϑ) are the classical action and oscillator phase. In the classical limit the expression (9.7) for the Hamiltonian takes the form

$$\mathcal{H}_{cl} = \alpha^*\alpha + \kappa\cos[(k/\sqrt{2})(\alpha^* + \alpha)] \sum_{n=-\infty}^{\infty} \delta(\tau - nT). \tag{9.11}$$

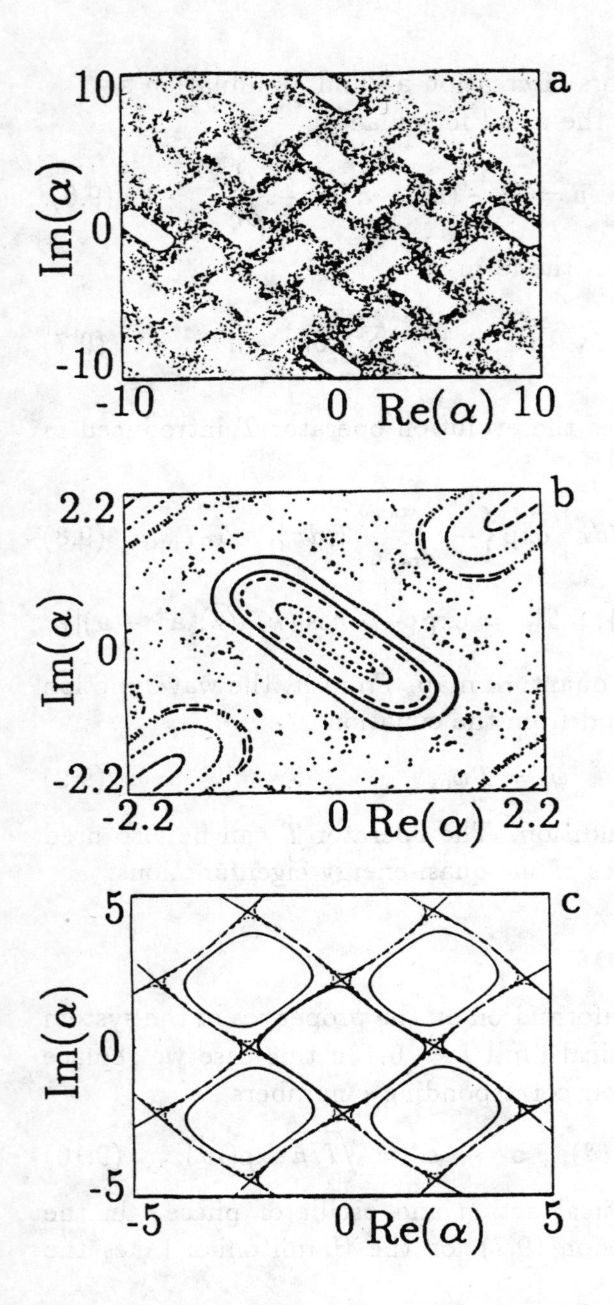

Fig. 10. Examples of the stochastic web generated by the classical map (9.12); (a): $K_H = 2; T = \pi/2; (r = 1; q = 4)$; (b): the same parameters as in (a), but different scale; (c): $K_H = 0.5$.

The map for the classical variables $\alpha_n \equiv \alpha(\tau_n - 0)$, $\alpha_n^* \equiv \alpha^*(\tau_n - 0)$ becomes

$$\alpha_{n+1} = e^{-iT}\{\alpha_n + i\kappa(k/\sqrt{2})\sin[(k/\sqrt{2})(\alpha_n^* + \alpha_n)]\}. \qquad (9.12)$$

The expression for α_{n+1}^* follows from (9.12) by complex conjugation. Though for convenience $\mathcal{H}_{cl}$ is normalized to $\hbar$, the Planck constant is absent in the finite expressions.

The analysis of equation (9.12) made in [55,56] (see also [69]) shows that for arbitrary values of the parameter

$$K_H = \kappa k^2 = \Omega_0^2 T_0/\omega_0, \quad (\Omega_0^2 = \varepsilon k_0^2/m_0 T_0), \qquad (9.13)$$

and under the condition of resonance (9.2) there exists stochastic dynamics of particles in certain regions of phase space. The frequency Ω_0 introduced in (9.13) refers to the frequency of small oscillations in the perturbed wave.

Regions of chaos exist at arbitrary small values of ε. In the phase space of the system they form a network of channels with symmetry of the order q which is called a stochastic web. The web for $q = 4$ is shown in Fig. 10a-c. Along the channels of the web there is unlimited transport of particles. This process, under the condition

$$\Delta I_n = |I_{n+1} - I_n| \ll 1 \qquad (9.14)$$

and $K_H \gg 1$, has the usual character of diffusion. According to (9.12) the inequality (9.14) means that

$$\Omega_0^2 T_0/k_0 v_0 = K_H(\omega_0/k_0 v_0) \ll 1, \qquad (9.15)$$

where v_0 is a characteristic particle velocity. Existence of the stochastic web at arbitrary values of ε must influence the dynamic properties of quantum particles in an essential way. Some of the peculiarities of quantum dynamics in this case will be discussed below.

9.4 Equation for the Generating Function

This rather formal section is devoted to constructing a recursion relation for the following generating function

$$\mathcal{R}_n(\gamma|\beta) = < \psi_n | e^{\gamma \mathbf{a}^+} e^{\beta \mathbf{a}} | \psi_n >, \qquad (9.16)$$

where the brackets $< \cdots >$ mean quantum mechanical average in the Schrödinger representation, and γ and β are arbitrary values. Using (9.16) one can calculate an arbitrary expectation value at discrete time n

$$< \psi_n|(\mathbf{a}^+)^l \mathbf{a}^s|\psi_n >= \frac{\partial^l}{\partial\gamma^l\partial\beta^s}\mathcal{R}(\gamma|\beta)|_{\gamma=\beta=0}. \tag{9.17}$$

Now we obtain a recurrence relation connecting functions $\mathcal{R}_n$ and $\mathcal{R}_{n-1}$. For this we use (9.3) and write (9.16) as

$$\mathcal{R}_n(\gamma|\beta) =< \psi_{n-1}|\hat{T}^+ e^{\gamma\mathbf{a}^+} e^{\beta\mathbf{a}}\hat{T}|\psi_{n-1} > . \tag{9.18}$$

By means of the explicit form of the operator $\hat{T}$ (9.8) and ordering of the operator functions (see Appendix B) we obtain for (9.16) the following recursion relation

$$\mathcal{R}_n(\gamma|\beta) = \sum_{m=-\infty}^{\infty} J_m\{2\kappa \sin[(k/\sqrt{2})(\beta e^{-iT} - \gamma e^{iT})]\} \tag{9.19}$$

$$\times \exp\left\{-\frac{m^2k^2}{4} + \frac{imk}{2\sqrt{2}}(\gamma e^{iT} + \beta e^{-iT})\right\}$$

$$\mathcal{R}_{n-1}(\gamma e^{iT} + imk/\sqrt{2}|\beta e^{-iT} + imk/\sqrt{2}),$$

where $J_m(z)$ is the Bessel function.

It is seen from (9.17) that expectation values without phase correlations can be obtained for $l = s \neq 0$. And, conversely, expectation values taking into account the phase correlations can be found from (9.16), e.g., at $\beta = -\gamma^*$, i.e. from the expression

$$\mathcal{R}(\gamma| -\gamma) =< \psi_n|e^{\gamma\mathbf{a}^+} e^{-\gamma\mathbf{a}}|\psi_n >= \mathcal{D}_n(\gamma)e^{|\gamma|^2}. \tag{9.20}$$

Using (9.19) and (9.20) we get a recursion relation for the function $\mathcal{D}_n(\gamma)$

$$\mathcal{D}_n(\gamma) = \sum_{m=-\infty}^{\infty} (-1)^m J_m\{2\kappa \sin[(k/\sqrt{2})(\gamma e^{iT} + \gamma^* e^{-iT})]\} \tag{9.21}$$

$$\mathcal{D}_{n-1}(\gamma e^{iT} + imk/\sqrt{2}).$$

Assuming that

$$\gamma = ik/\sqrt{2}, \tag{9.22}$$

we obtain from (9.20) and the definitions (9.4) and (9.6)

$$\mathcal{D}_n(ik/\sqrt{2}) = <\psi_n|e^{ik_0x}|\psi_n> . \tag{9.23}$$

Expression (9.23) corresponds to the expectation value of the simplest correlator characterizing phase properties of the wave packet.

Now we derive the classical limit of (9.21). For this we should take into account that according to (9.4), (9.6) and (9.9) we have for $k, \kappa, \mathbf{a}, \mathbf{a}^+$ and γ

$$\gamma \sim k \sim \sqrt{\hbar}, \quad \kappa \sim 1/\hbar, \quad \mathbf{a} \sim \mathbf{a}^+ \sim \sqrt{I/\hbar}.$$

Using this and (9.21) we find at $\hbar \to 0$

$$\mathcal{D}_n^{cl}(\gamma) = \sum_{m=-\infty}^{\infty} (-1)^m J_m\{(\kappa k/\sqrt{2})(\gamma e^{iT} + (\gamma^* e^{-iT})\} \tag{9.24}$$

$$\mathcal{D}_{n-1}^{cl}(\gamma e^{iT} + imk/\sqrt{2}).$$

Below we shall give a quasiclassical estimate of the correlator $\mathcal{D}_n(ik/\sqrt{2})$ (9.23) and compare it with the classical limit.

9.5 The Region of Parameters for Quasiclassical Consideration

To write down the explicit expression (9.23) with the use of the recursion formula (9.21) we shall introduce the following notations

$$\gamma_0 = \gamma = ik/\sqrt{2}, \tag{9.25}$$

$$\gamma_1 = \gamma_0 e^{iT} + im_1 k/\sqrt{2} = i(k/\sqrt{2})(e^{iT} + m_1),$$

$$\dots$$

$$\gamma_l = \gamma_{l-1} e^{iT} + im_l k/\sqrt{2} = i(k/\sqrt{2})(e^{ilT} + m_1 e^{i(l-1)T} + \dots + m_l),$$
$$(l = 1, ..., n),$$

$$\xi_1 = -(k/2\sqrt{2})(\gamma_1 + \gamma_1^*) = (k^2/2)\sin T, \tag{9.26}$$

$$\cdots$$

$$\xi_l = -(k/2\sqrt{2})(\gamma_l + \gamma_l^*) = (k^2/2)[\sin(lT) + m_1 \sin(l-1)T$$
$$+ \cdots + m_{l-1}\sin T], \quad (l = 1, ..., n; m_0 = 0).$$

Taking into account (9.21), (9.22), (9.25) and (9.26) expression (5.23) can be rewritten as

$$\mathcal{D}_n(ik/\sqrt{2}) = \sum_{m_1,...,m_n=-\infty}^{\infty} J_{m_1}(2\kappa \sin\xi_1) J_{m_2}(2\kappa\sin\xi_2)\cdots \tag{9.27}$$
$$J_{m_n}(2\kappa\sin\xi_n)\mathcal{D}_0(\gamma_n),$$

where $\mathcal{D}_0(\gamma_n)$ is the initial condition. According to (9.20) this can be represented as

$$\mathcal{D}_0(\gamma_n) = < \psi_0 | e^{\gamma_n \mathbf{a}^+} e^{-\gamma_n^* \mathbf{a}} | \psi_0 > e^{-|\gamma_n|^2/2}, \tag{9.28}$$

where ψ_0 is an arbitrary initial state. Operators $\mathbf{a}$ and $\mathbf{a}^+$ in (9.28) are determined at the initial time $n = 0$, and their action on ψ_0 is known.

In the classical limit it is necessary to replace $\sin\xi_l$ in (9.27) by ξ_l and the expression (9.27) will be as follows (see also (9.24))

$$\mathcal{D}_n^{cl}(ik/\sqrt{2}) = \sum_{m_1,...,m_n=-\infty}^{\infty} J_{m_1}\{K_H \sin T\} J_{m_2}\{K_H[\sin(2T) + m_1 \sin T]\}$$
$$\cdots J_{m_n}\{K_H[\sin(nT) \tag{9.29}$$
$$+ m_1 \sin(n-1)T + \cdots$$
$$+ m_{n-1}\sin T]\}\mathcal{D}_0^{cl}(\gamma_n).$$

The initial condition $\mathcal{D}_0^{cl}(\gamma_n)$ in the classical limit is determined by the following expression

$$\mathcal{D}_0^{cl} = \frac{1}{\pi}\int d^2\alpha P(\alpha^*, \alpha) e^{\gamma_n \alpha^* - \gamma_n^* \alpha}, \quad (d^2\alpha = d(Re\alpha)d(Im\alpha)), \tag{9.30}$$

where $P(\alpha^*, \alpha)$ is the initial distribution function (see also (5.29), (5.35)). According to (9.6) $\mathcal{D}_0^{cl}$ can be expressed through the initial action and phase. Moreover, comparing quantum (9.27) and classical (9.29) expressions we need an additional requirement concerning the transformation of the initial condition (9.28) at $\hbar \to 0$ to the classical initial condition (9.30). This correspondence is known to be realizable by various methods, and in what follows we shall consider the requirement to be satisfied.

It is seen from (9.27) and (9.26) that along with the initial condition (9.28) quantum dynamics is determined by three independent parameters k, κ, T. It is convenient to choose the independent parameters

$$K_H = \kappa k^2, \quad \kappa, \quad T, \tag{9.31}$$

where T and K_H are classical parameters, and κ is a quantum parameter. From (9.12) it follows that we can estimate the number of quanta absorbed by the oscillator during one kick as

$$\Delta I/\hbar \sim \kappa\sqrt{Ik_0^2/\omega_0 m_0}, \tag{9.32}$$

where I is the characteristic action. The quasiclassical condition for the perturbation is $\Delta I/\hbar \gg 1$.

Now let us introduce a "quantum boundary of stochasticity" (QBS) in the same way as done in section 6.4. The QBS separates the region of parameters (9.31) with purely quantum dynamics, from the region of parameters where at least under the finite time-interval $n_\hbar > n > 1$ the quantum dynamics can be approximated by the classical dynamics. It is seen from (9.27) that the dynamics is essentially quantum at times t larger than T_0 under the condition $|\xi_1| \sim 1$, which according to (9.26) gives

$$\frac{k^2}{2}|\sin T| = \frac{K_H}{2\kappa}|\sin T| \sim 1. \tag{9.33}$$

In the region of parameters (9.31), where the inverse inequality

$$\frac{K_H}{2\kappa}|\sin T| \ll 1 \tag{9.34}$$

is satisfied, the dynamics of the quantum system approximately coincides with that of the classical system. The characteristic time-scale $n_\hbar$ of quasiclassical description of the system (9.1) will be estimated below.

9.6 Time-Scale of Quasiclassical Description for Strong Chaos

Earlier we introduced a quantum correlator $\mathcal{D}_n$ in the formula (9.27) and its classical limit $\mathcal{D}_n^{cl}$ in formula (9.29). Comparison of both expressions allows us to find when the quantum dynamics approximates its classical counterpart. The convenience of this approach is due to the fact that we analyse and compare not the finite solutions but their formal iteration representations. Such an approach makes the solution of the problem much simpler.

Consider first the case of strong chaos (in the classical limit), i.e. inequality $K_H \gg 1$. The applicability of the classical expression (9.29) for describing the quantum dynamics at finite times $n \gg 1$ means that in (9.27) at least the condition

$$|\xi_l| \ll 1, \quad (l = 1, ..., n) \tag{9.35}$$

should be satisfied, and all functions $\sin \xi_l$ in (9.27) may be approximately replaced by ξ_l. Assuming (9.35) to be satisfied and taking into account that for

$$|m_l| \gg 2\kappa|\xi_l| = K_H|\sin(lT) + m_1 \sin(l-1)T + \cdots + m_{l-1}\sin T|, \tag{9.36}$$

the Bessel functions $J_{m_l}(2\kappa \sin \xi_l)$ are exponentially small, we have from (9.36) an estimate for the characteristic number of terms m_l $(l = 1, ..., n)$ in (9.27)

$$|m_1| \sim 2\kappa|\xi_1| = K_H|\sin T|, \tag{9.37}$$

$$|m_2| \sim 2\kappa|\xi_2| = K_H|\sin(2T) + m_1 \sin T|,$$

$$\cdots$$

$$|m_n| \sim 2\kappa|\xi_n| = K_H|\sin(nT) + m_1 \sin(n-1)T + \cdots + m_{n-1}\sin T|.$$

It is clear from (9.37) that the estimate of the number of terms in (9.27) as well as of the quantum dynamics of the correlator $\mathcal{D}_n(ik/\sqrt{2})$ mostly depends not only on the parameter K_H, but also on the value $\sin T$. So, we shall consider two limiting cases, strong chaos

$$K = K_H|\sin T| \gg 1, \tag{9.38}$$

and weak chaos

$$K \ll 1. \tag{9.39}$$

First let us discuss strong chaos (9.38). From (9.37) we have

$$|m_1| \sim K_H|\sin T| = K \gg 1, \tag{9.40}$$

$$|m_2| \sim K_H|\sin T||m_1| \sim K^2,$$

$$\ldots$$

$$|m_n| \sim K_H|\sin T||m_{n-1}| \sim K^n = \exp(n \ln K).$$

From (9.37), (9.40) one can estimate the value $|\xi_n|$

$$|\xi_n| \sim \frac{|m_n|}{2\kappa} \sim \frac{\exp(n \ln K)}{2\kappa}. \tag{9.41}$$

From condition (9.35) and taking account of (9.41) one can estimate the time of the quasiclassical description $n_\hbar$ (i.e. the time of quasiclassical approach)

$$n < n_\hbar = \frac{\ln(2\kappa)}{\ln K}, \quad (K \gg 1). \tag{9.42}$$

Thus, in the region of parameters (9.34) the dynamics of the quantum system at finite times n approximately coincides with dynamics of the classical system, and under the additional condition (9.38) the characteristic time-scale $n_\hbar$ of the classical description has the order (9.42), that is in agreement with the estimate (1.2). Note, that under the condition (9.38) and the additional conditions $K_H \gg 1$ and $|\sin T| \sim 1$ resonances (9.2) with $q = 3 - 7$ are possible, which were studied in detail in the classical limit in [55,56,69]. In this case the motion of a classical particle in the stochastic region of phase space is diffusive and the stochastic region is characterized by different symmetries depending on q. The dynamics of a classical particle in the stochastic web is discussed in detail also in [70]. Some properties of the symmetry of the eigenfunctions in this case are discussed in [68]. Now we shall discuss the case of weak chaos (9.39).

9.7 Quantum Dynamics in the Region of Weak Chaos

Under the condition $K \ll 1$ the classical dynamics is regular in most regions of phase space, and dynamical chaos is realized in some comparatively small regions. As was already mentioned, under the resonant condition (9.2) these regions create a thin stochastic web with a symmetry of the order q [55,56,69]. A quantum treatment of these thin stochastic regions is interesting, so far as quantum effects in these regions can significantly influence the characteristic properties of particle dynamics. Here we will be mainly interested in an initial population of the system (9.1) in the region with chaotic component under the condition of weak chaos $K \ll 1$. This section is divided into two parts (9.7.1, and 9.7.2). In the first part we consider the quantum effects connected only with the dynamics of the quantum map (9.27) under condition (9.39), without taking into account the initial conditions. We show that characteristic time-scale $n_\hbar$ connected with these "quantum map"- effects is of the order $n_\hbar \sim 1/\hbar$. In the second part of this section we analyse the quantum effects connected with the initial population of wave packet. We show that the initial population of the wave packet in the unstable regions of classical phase space can lead to an appearance of logarithmically small time-scale of the order (1.2), if the initial packet is rather narrow ($\Delta p \Delta x \sim \hbar$). In the case of rather wide initial packet ($\Delta p \Delta x \gg \hbar$), the time-scale $n_\hbar$ is of the order: $n_\hbar \sim 1/\hbar$.

9.7.1 Time-Scale $n_\hbar$ for Quantum Map

When analysing the dynamics of the correlator (9.27) under the weak chaos condition $K \ll 1$ we deal with two limiting cases,

$$K_H \ll 1, \quad \sin T \leq 1, \tag{9.43}$$

$$K_H \sim 1, \quad \sin T \ll 1. \tag{9.44}$$

First, consider the case (9.43). For the sake of simplicity we shall confine ourselves to the resonance $T = \pi/2$ ($q = 4$). The classical phase space in this case is shown in Fig. 10a-c. From (9.26) we have

$$\xi_1 = K_H/2\kappa, \tag{9.45}$$

$$\xi_2 = (K_H/2\kappa)m_1,$$

$$\xi_3 = (K_H/2\kappa)(-1 + m_2),$$

$$\xi_4 = (K_H/2\kappa)(-m_1 + m_3),$$

$$\dots$$

$$\xi_n = (K_H/2\kappa)[\sin(n\pi/2) + m_1 \sin(n-1)\pi/2 + \cdots + m_{n-1}].$$

Suppose the condition $|\xi_l| \ll 1$ holds which, according to the results of section 9.5, is necessary for the classical description to be valid at finite times. From (9.27), taking into account (9.43) and (9.45) we estimate the maximum $m_l (l = 1, ..., n)$ in (9.27)

$$|m_1| \sim 2\kappa|xi_1| = K_H \ll 1, \tag{9.46}$$

$$|m_2| \sim 2\kappa|\xi_2| \sim K_H^2,$$

$$|m_3| \sim 2\kappa|\xi_3| \sim K_H + K_H^3,$$

$$|m_4| \sim 2\kappa|\xi_4| \sim K_H^2 + K_H^5,$$

$$\dots$$

$$|m_{2n-1}| \sim K_H + (2n-3)K_H^3 + O(K_H^5); \quad (n = 2, ...),$$

$$|m_{2n}| \sim nK_H^2 + O(K_H^4), \quad (n = 1, ...).$$

The estimate of the value $|\xi_n|$ from (9.46) gives

$$|\xi_m| \sim \begin{cases} nK_H^3/\kappa, & (m = 2n-1), \\ nK_H^2/2\kappa, & (m = 2n). \end{cases} \tag{9.47}$$

From the condition $|\xi_m| \ll 1$ we obtain that the time-scale $n_\hbar$ of the classical description of the correlator dynamics (9.27) is

$$n < n_\hbar^{(1)} = 2\kappa/K_H^2. \tag{9.48}$$

As was already mentioned before, the time-scale (9.48) takes into account only the action of quantum map (9.27). According to (9.48) this time-scale appears to be of the order $1/\hbar$, and essentially exceeds the logarithmically small estimate (9.42) for the case of strong chaos. The influence of the initial population will be considered in section 9.7.2.

Now let us analyse the case (9.44). From (9.26) we have

$$\xi_1 \sim K/2\kappa \ll 1, \quad |m_1| \sim K \ll 1, \tag{9.49}$$

$$\xi_2 \sim (K/2\kappa)[2 + O(K)], \quad |m_2| \sim 2K + O(K^2),$$

$$\cdots$$

$$\xi_n \sim (K/2\kappa)[n + O(K)], \quad |m_n| \sim nK + O(K^2), \quad (nT \ll 1).$$

We require that the condition of the classical description on times n be satisfied, which leads to the inequality

$$\xi_n \sim nK/2\kappa \ll 1. \tag{9.50}$$

Then we have

$$n < n_{\hbar}^{(2)} = 2\kappa/K. \tag{9.51}$$

However, one should bear in mind that the estimates (9.49) are also obtained under the condition $nT \ll 1$, therefore the time of the classical description in the case (9.44) should be estimated based on

$$n_{\hbar}^{(2)} = min(2\kappa/K, 1/T). \tag{9.52}$$

From (9.52) it follows that in the quasiclassical region

$$2\kappa/K \gg 1/T \gg 1, \tag{9.53}$$

the quasiclassical description of the dynamics of the quantum system is limited by the time $n_{\hbar}^{(2)} \sim 1/T$. When the quasiclassical region is not very deep

$$1/T \gg 2\kappa/K \gg 1, \tag{9.54}$$

the time $n_{\hbar}^{(2)}$ is evaluated by (9.51).

The case (9.44) is apparently of no interest since the condition $nT \ll 1$ shows that the time under consideration is less than a period of oscillation

$$t = nT_0 \ll 1/\omega_0.$$

In contrast, the case (9.43) is of great interest. Now we consider the influence on the time-scale $n_{\hbar}$ of the initial population of the system.

9.7.2 The Influence of the Initial Conditions

Here we are considering only the case of weak chaos (9.43). As was already mentioned, the classical dynamics in this case is regular in the most regions of phase space. So, the initial population of quantum system (9.1) in these regions leads to the time-scale $n_{\hbar}^{(1)} \sim 1/\hbar$ (9.48) of the applicability of classical dynamics. Let us consider the case when at $t = 0$ the wave packet is populated in the unstable region, and moreover, in the vicinity of the unstable (hyperbolic) point (see Fig. 11).

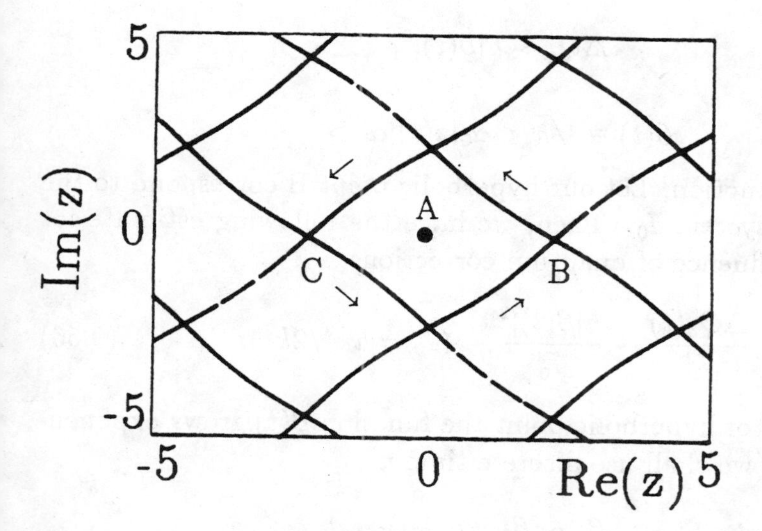

Fig. 11. The classical phase space (z^*, z) of the map (9.12); A-elliptic point; B,C-hyperbolic points. By arrows the action of map (9.12) is shown, beginning in the vicinity of the hyperbolic point B.

In this case, as the results derived in section 9.7.1 show, the *dynamics* of the quantum map (9.21) leads to the time-scale $n_{\hbar} \sim 1/\hbar$. So, we investigate below the influence of quantum effects on the initial population of the wave packet under the condition of weak chaos (9.43). The simplest (qualitative) way to estimate these effects can be done using the following arguments. Assume, that we put a quantum particle in the vicinity of the hyperbolic point (for example, the point B in Fig. 11), in CS $|\alpha_0>$, which occupies the cell in phase space $\Delta p \Delta x \sim \hbar$ (here p and x are dimensional variables). Let us consider the time-evolution of the operator, for example, $\hbar^2(\mathbf{a}^+\mathbf{a})^2$. The time-dependent expectation value of this operator is given by the expression

$$Q(t) \equiv \hbar^2 < \alpha_0| \left(\mathbf{a}^+(t)\mathbf{a}(t)\right)^2 |\alpha_0> = \hbar^2 < \alpha_0|\mathbf{a}^{+2}(t)\mathbf{a}^2(t)|\alpha_0> \tag{9.55}$$
$$+\hbar^2 < \alpha_0|\mathbf{a}^+(t)\mathbf{a}(t)|\alpha_0> .$$

The contribution of quantum effects in (9.55) can be estimated by the expression

$$\Delta Q_q \sim \hbar|\beta(t)|^2,$$

where

$$\beta(t) \equiv \sqrt{\hbar} < \alpha_0|\mathbf{a}(t)|\alpha_0>$$

is a classical function. Let our hyperbolic point B correspond to the action of the system I_0. Then, we have the following estimate for the relative influence of quantum corrections

$$\frac{\Delta Q_q(t)}{I_0^2} \sim \frac{\hbar|\beta(t)|^2}{I_0^2}, \quad (I_0 = \hbar\pi^2/2k^2). \tag{9.56}$$

In the vicinity of hyperbolic point the function $\beta(t)$ grows exponentially in time (we shall use discrete time n),

$$\beta_n \sim \beta_0 \exp(\lambda n).$$

Then, under the condition $\Delta Q_q/I_0^2 \sim 1$, we derive from (9.56) the time-scale $n_\hbar$ of applicability of the classical approach

$$n_\hbar \sim \frac{1}{2\lambda} \ln\left(\frac{I_0^2}{\hbar|\beta_0|^2}\right), \tag{9.57}$$

where $\lambda \approx 2K_H \ll 1$ is a characteristic increment of classical instability.

This estimate still doesn't take into account the peculiarities of the initial distribution (for example, the width of the initial packet). We shall analyze this problem by using the projection formulas for operator functions on the c-number phase space [58]. Following paper [58], we represent an arbitrary operator function $\mathbf{G}(\mathbf{a}^+, \mathbf{a})$ in the form

$$\mathbf{G}(\mathbf{a}^+, \mathbf{a}) = \int d^2 z F(z, z^*) \hat{\Delta}^{(\Omega)}(z - \mathbf{a}, z^* - \mathbf{a}^+), \tag{9.58}$$

where $F(z, z^*)$ is a c-number equivalent of the operator function $\mathbf{G}(\mathbf{a}^+, \mathbf{a})$, and the operator function $\hat{\Delta}^{(\Omega)}$ has the form

$$\hat{\Delta}^{(\Omega)}(z-\mathbf{a}, z^*-\mathbf{a}^+) = \frac{1}{\pi^2} \int d^2\alpha \Omega(\alpha^*, \alpha) \exp\{-[\alpha(z^*-\mathbf{a}^+) - \alpha^*(z-\mathbf{a})]\}. \tag{9.59}$$

The c-number equivalent $F(z, z^*)$ has the following representation

$$F(z, z^*) = \pi Tr[\mathbf{G}(\mathbf{a}^+, \mathbf{a}) \hat{\Delta}^{(\tilde{\Omega})}(z - \mathbf{a}, z^* - \mathbf{a}^+)], \tag{9.60}$$

where

$$\tilde{\Omega}(\alpha^*, \alpha) = \frac{1}{\Omega(-\alpha^*, -\alpha)}. \tag{9.61}$$

The function $F(z, z^*)$ is called also c-number projection of the operator $\mathbf{G}(\mathbf{a}^+, \mathbf{a})$ on the phase space (z^*, z). The choice of the function $\Omega(\alpha^*, \alpha)$ defines the representation. In particular, $\Omega(\alpha^*, \alpha) = 1$ corresponds to the Weyl transformation of the operator, $F^{(W)}(z, z^*)$. We shall deal below, in this section, only with the Weyl transformation. The expectation value of an arbitrary operator $\mathbf{G}(\mathbf{a}^+, \mathbf{a})$ can be expressed through the projections of this operator $F(z, z^*)$ and density matrix $\rho^{(W)}(z, z^*)$ on the phase space. Namely,

$$< \mathbf{G} > \equiv Tr\,(\mathbf{G}\hat{\rho}) = \frac{1}{\pi} \int d^2 z F^{(W)}(z, z^*) \rho^{(W)}(z, z^*). \tag{9.62}$$

The projection $\rho^{(W)}(z, z^*)$ of the density matrix $\hat{\rho}$ on the phase space (z, z^*) is called the Wigner function. The analysis of Wigner function dynamics for classically chaotic system is given in section 10. In this section we shall use the Wigner function to analyse the influence of initial conditions on the time-scale $\tau_\hbar$ in the region of weak chaos in the stochastic web.

From (9.62) it follows, that to study the dynamics of expectation values of arbitrary operator $\mathbf{G}$, we can use the equation

$$< \mathbf{G}_n >= \frac{1}{\pi} \int d^2 z F^{(W)}(z, z^*) \rho_n^{(W)}(z, z^*), \tag{9.63}$$

where the projection $F^{(W)}(z, z^*)$ of the operator $\mathbf{G}$ remains time-independent, and the Wigner function $\rho_n^{(W)}(z, z^*)$ depends on time n. So, the first problem which we consider in this section is connected with the dynamics of the Wigner function $\rho_n^{(W)}(z, z^*)$ under the action of the evolution operator $\hat{T}$ (9.8). In Fig. 11 the classical phase space (z, z^*) is shown (A: $z = 0$ corresponds to elliptic point, and B: $z = \pi/\sqrt{2}k$, C: $z = -\pi/\sqrt{2}k$ correspond to the hyperbolic points). We shall limit ourselves by investigating the dynamics in a small vicinity of the points A, B, C, in Fig. 11. First, we study the evolution of the Wigner function by the action of the kick operator $\hat{T}_{int}$ in (9.8). For this, we expand the operator $\hat{T}_{int}$ in the vicinity of these points in a power series. For example, in the vicinity of the point A we have

$$\cos[(k/\sqrt{2})(\mathbf{a}^+ + \mathbf{a})] \approx 1 - \frac{k^2}{4}(\mathbf{a}^+ + \mathbf{a})^2. \tag{9.64}$$

Such an expansion of an operator function in the vicinity of the point A can be justified by the fact that the transformation to the c-number projections $\mathbf{a}^+ \to z^*$; $\mathbf{a} \to z$ makes the neglected terms in (9.64) small. In this case we have

$$\hat{T}_{int} = \exp\{-i\kappa \cos[(k/\sqrt{2})(\mathbf{a}^+ + \mathbf{a})]\} \approx e^{i\lambda(\mathbf{a}^+ + \mathbf{a})^2}, \tag{9.65}$$

$$(\lambda = \kappa k^2/4 = K_H/4),$$

where we have neglected a constant phase. Using the operator equality

$$\hat{T}_{int} e^{\alpha \mathbf{a}^+ - \alpha^* \mathbf{a}} \hat{T}_{int} = e^{-i\lambda(\mathbf{a}^+ + \mathbf{a})^2} e^{\alpha \mathbf{a}^+ - \alpha^* \mathbf{a}} e^{i\lambda(\mathbf{a}^+ + \mathbf{a})^2} \tag{9.66}$$

$$= \exp\{[\alpha - 2i\lambda(\alpha^* + \alpha)]\mathbf{a}^+ - [\alpha^* + 2i\lambda(\alpha^* + \alpha)]\mathbf{a}\},$$

and (9.59), we have

$$\hat{T}_{int}^+ \hat{\Delta}^{(1)}(z - \mathbf{a}, z^* - \mathbf{a}^+) \hat{T}_{int} \tag{9.67}$$

$$= \frac{1}{\pi^2} \int d^2\alpha e^{-\alpha z^* + \alpha^* z} \exp\{[\alpha - 2i\lambda(\alpha^* + \alpha)]\mathbf{a}^+$$

$$-[\alpha^* + 2i\lambda(\alpha^* + \alpha)]\mathbf{a}\}.$$

Substitution

$$\gamma = \alpha - 2i\lambda(\alpha^* + \alpha)$$

in (9.67) gives

$$\hat{T}^+_{int}\hat{\Delta}^{(1)}(z - \mathbf{a}, z^* - \mathbf{a}^+)\hat{T}_{int} \tag{9.68}$$

$$= \frac{1}{\pi^2} \int d^2\gamma \exp\left\{\gamma^*[z - 2i\lambda(z + z^*) - \mathbf{a}] - \gamma[z^* + 2i\lambda(z + z^*) - \mathbf{a}^+]\right\}$$

$$= \hat{\Delta}^{(1)}\left(z - 2i\lambda(z + z^*) - \mathbf{a}, z^* + 2i\lambda(z + z^*) - \mathbf{a}^+\right).$$

Using (9.68), we get from (9.60) the Wigner function after an interaction (kick)

$$\rho^{(W)}_{T_{int}}(z, z^*) = \pi Tr\left\{\hat{T}_{int}\hat{\rho}\hat{T}^+_{int}\hat{\Delta}^{(1)}(z - \mathbf{a}, z^* - \mathbf{a}^+)\right\} \tag{9.69}$$

$$= \pi Tr\left\{\hat{\rho}\hat{T}^+_{int}\hat{\Delta}^{(1)}(z - \mathbf{a}, z^* - \mathbf{a}^+)\hat{T}_{int}\right\}$$

$$= \rho^{(W)}[z - 2i\lambda(z^* + z), z^* + 2i\lambda(z^* + z)].$$

The equation (9.69) describes the evolution of the Wigner function in the vicinity of the point A in Fig. 11 by one kick. Similar equations may be derived in the vicinity of the points B and C. For example, in the vicinity of the point B we have

$$\cos[(k/\sqrt{2})(\mathbf{a}^+ + \mathbf{a})] = \cos[(k/\sqrt{2})(\mathbf{b}^+ + \mathbf{b})], \tag{9.70}$$

where the following substitution is made

$$\mathbf{b} = \mathbf{a} - \pi/\sqrt{2}k. \tag{9.71}$$

Thus, the evolution operator $\hat{T}_{int}$ in the vicinity of the point B has the form

$$\hat{T}_{int} = \exp\{i\kappa \cos[(k/\sqrt{2})(\mathbf{b}^+ + \mathbf{b})]\} \approx e^{-i\lambda(\mathbf{b}^+ + \mathbf{b})^2}, \tag{9.72}$$

and differs from (9.65) only by the sign in the exponent. So, we find analogously to (9.69)

$$\rho^{(W)}_{T_{int},b}(z,z^*) = \rho^{(W)}_b[z + 2i\lambda(z^* + z), z^* - 2i\lambda(z^* + z)], \tag{9.73}$$

where

$$\rho^{(W)}_b(z,z^*) \equiv \pi Tr\left\{\hat{\rho}\hat{\Delta}^{(1)}(z - \mathbf{b}, z^* - \mathbf{b}^+)\right\}. \tag{9.74}$$

The relation between $\rho^{(W)}(z,z^*)$ and $\rho^{(W)}_b(z,z^*)$ is given by the following expression

$$\begin{aligned}\rho^{(W)}_b(z,z^*) &= \pi Tr\left\{\hat{\rho}\hat{\Delta}^{(1)}(z + \pi/\sqrt{2}k - \mathbf{a}, z^* + \pi/\sqrt{2}k - \mathbf{a}^+)\right\} \\ &= \rho^{(W)}(z + \pi/\sqrt{2}k, z^* + \pi/\sqrt{2}k).\end{aligned} \tag{9.75}$$

Finally, we derive the equation which describes the evolution of the Wigner function by the kick, in the vicinity of the point B,

$$\begin{aligned}\rho^{(W)}_{T_{int}}(z,z^*) &= \rho^{(W)}_{T_{int},b}(z - \pi/\sqrt{2}k, z^* - \pi/\sqrt{2}k) \\ &= \rho^{(W)}_b[z - \pi/\sqrt{2}k + 2i\lambda(z^* + z - \sqrt{2}\pi/k), C.C.] \\ &= \rho^{(W)}[z + 2i\lambda(z^* + z - \sqrt{2}\pi/k), C.C.].\end{aligned} \tag{9.76}$$

Analogously, in the vicinity of the point C we have

$$\rho^{(W)}_{T_{int}}(z,z^*) = \rho^{(W)}[z + 2i\lambda(z^* + z + \sqrt{2}\pi/k), C.C.]. \tag{9.77}$$

Now we present the evolution of Wigner function by action of the operator $\hat{T}_0$ in (9.8). Simple calculations give

$$\begin{aligned}\rho^{(W)}_{T_0}(z,z^*) &= \pi Tr\left\{\hat{T}_0\hat{\rho}\hat{T}^+_0\hat{\Delta}^{(1)}(z - \mathbf{a}, z^* - \mathbf{a}^+)\right\} \\ &= \rho(ze^{iT}, z^*e^{-iT}).\end{aligned} \tag{9.78}$$

Finally, we have the following formulas for the evolution of the Wigner function $\rho^{(W)}(z,z^*)$ by the operator $\hat{T}$ (9.8), starting from the points A, B, C.

From the point A

$$\rho^{(W)}_T(z,z^*) = \rho^{(W)}[e^{iT}(z - 2i\lambda(z^* + z)), C.C.] \equiv \rho^{(W)}(z', z'^*). \tag{9.79}$$

From the point B

$$\rho_T^{(W)}(z,z^*) \tag{9.80}$$
$$= \rho^{(W)}[e^{iT}\left((z+2i\lambda(z^*+z-\sqrt{2}\pi/k)\right), C.C.] \equiv \rho^{(W)}(z',z'^*).$$

From the point C

$$\rho_T^{(W)}(z,z^*) = \rho^{(W)}[e^{iT}\left(z+2i\lambda(z^*+z+\sqrt{2}\pi/k)\right), C.C.] \tag{9.81}$$
$$\equiv \rho^{(W)}(z',z'^*).$$

For simplicity, we consider here only the case $T = \pi/2$, shown in Fig. 10a-c. In this case the problem reduces to the trivial map shown schematically in Fig. 11.

Consider the motion which starts in the vicinity of the hyperbolic point B, and under the condition of weak chaos in the classical limit

$$\lambda = \frac{\kappa k^2}{4} = \frac{K_H}{4} \ll 1. \tag{9.82}$$

From (9.79)-(9.81) it follows that for the cycle of four transformations we again arrive at the vicinity of the point B. This cycle-transformation can be presented in the following simple form

$$\xi' = \xi + 8i\lambda\xi^*, \tag{9.83}$$

where ξ is a local complex coordinate in the vicinity of the point B

$$\xi = z - \frac{\pi}{\sqrt{2}k}. \tag{9.84}$$

Introducing $\xi = x + iy$, we represent (9.83) in the following form

$$\begin{pmatrix} x \\ y \end{pmatrix}' = \begin{pmatrix} 1 & -8\lambda \\ -8\lambda & 1 \end{pmatrix}\begin{pmatrix} x \\ y \end{pmatrix}. \tag{9.85}$$

From (9.85) we derive the eigenvectors and the eigenvalues

$$|1> = \frac{1}{\sqrt{2}}\begin{pmatrix} 1 \\ -1 \end{pmatrix}, \quad |2> = \frac{1}{\sqrt{2}}\begin{pmatrix} 1 \\ 1 \end{pmatrix}, \tag{9.86}$$
$$\epsilon_1 = 1 - 8\lambda, \quad \epsilon_2 = 1 + 8\lambda.$$

It is clear that the initial packet will be stretched by the transformation (9.83).

To investigate the role of quantum effects in this "stretching mechanism", we represent the initial density matrix in the form (5.29), (5.35), which is convenient for comparison with the classical consideration, and calculate the time-evolution of the quantum average value of the operator **G** (9.58). As the operator **G** we choose, for example, the following one

$$\mathbf{G} = e^{i\gamma_0\sqrt{\hbar}(\mathbf{a}^+ + \mathbf{a})}. \tag{9.87}$$

The Weyl projection $F^{(W)}(\beta, \beta^*)$ of the operator (9.87) on the phase space (β^*, β) (where $\beta = \sqrt{\hbar} z$ is a dimensional variable), is

$$F^{(W)}(\beta, \beta^*) = e^{i\gamma_0(\beta+\beta^*)}. \tag{9.88}$$

The initial density matrix $\hat{\rho}_0$ (5.29), (5.35) has, according to (9.60), the following Weyl projection $\rho_0^{(W)}(\beta, \beta^*)$ on the phase space (β, β^*)

$$\rho_0^{(W)}(\beta, \beta^*) = \frac{1}{(\sigma^2 + \hbar/2)} e^{-\frac{1}{(\sigma^2+\hbar/2)}|\beta-\beta_0|^2}, \quad (\sigma^2 = 1/\nu), \tag{9.89}$$

where σ and α_0 are parameters which characterize the width and the center of the initial distribution.

From (9.84) we have a local dimensional coordinate in the vicinity of the point B: $\xi = \beta - \pi\sqrt{\hbar}/\sqrt{2}k$, for which we shall use the same letter as in (9.84). Then, for (ξ, ξ^*) we have the map, which describes the dynamics in the vicinity of the point B (see (9.83)), in the linear approximation

$$\begin{pmatrix} \xi \\ \xi^* \end{pmatrix}' = \begin{pmatrix} 1 & i8\lambda \\ -8i\lambda & 1 \end{pmatrix} \begin{pmatrix} \xi \\ \xi^* \end{pmatrix}. \tag{9.90}$$

The iteration of (9.90) gives the coordinates $(\xi, \xi^*)_n$ at the time n

$$\begin{pmatrix} \xi \\ \xi^* \end{pmatrix}_n = \begin{pmatrix} \frac{1}{2}(\epsilon_1^n + \epsilon_2^n) & \frac{i}{2}(\epsilon_2^n - \epsilon_1^n) \\ \frac{1}{2}(\epsilon_1^n - \epsilon_2^n) & \frac{1}{2}(\epsilon_1^n + \epsilon_2^n) \end{pmatrix} \begin{pmatrix} \xi \\ \xi^* \end{pmatrix}. \tag{9.91}$$

Using (9.80), (9.91), we derive from (9.63) an expression which describes the evolution of the average value of the operator (9.87)

$$< \mathbf{G}_n >= \frac{1}{\pi(\sigma^2+\hbar/2)} e^{i\gamma_0\pi\sqrt{2\hbar}/k} \int d^2\xi e^{i\gamma_0(\xi+\xi^*)} \tag{9.92}$$

$$\times \exp\left\{ -\frac{1}{(\sigma^2+\hbar/2)} \left| \frac{1}{2}(\epsilon_1^n+\epsilon_2^n)\xi + \frac{i}{2}(\epsilon_2^n-\epsilon_1^n)\xi^* - \tilde{\beta}_0 \right|^2 \right\}.$$

In (9.92) $\tilde{\beta}_0 = \beta_0 - \pi\sqrt{\hbar}/\sqrt{2}k$. Integration of (9.92) gives

$$< \mathbf{G}_n >= f(n) \exp\left\{ -\frac{\gamma_0^2}{2}(\sigma^2+\frac{\hbar}{2}) \left[\frac{\epsilon_1^{2n}+\epsilon_2^{2n}}{(\epsilon_1\epsilon_2)^{2n}} \right] \right\}, \tag{9.93}$$

$$f(n) = (\epsilon_1\epsilon_2)^n e^{i\gamma_0\pi\sqrt{2\hbar}/k} \exp\left\{ 2\gamma_0\tilde{\beta}_0 \frac{\epsilon_2^n-\epsilon_1^n}{\epsilon_1^{2n}+\epsilon_2^{2n}} + 2i\gamma_0\tilde{\beta}_0 0 \frac{\epsilon_1^n+\epsilon_2^n}{\epsilon_1^{2n}+\epsilon_2^{2n}} \right\}.$$

For large n we have from (9.93)

$$< \mathbf{G}_n >\sim \gamma(n) \exp\left\{ -\frac{\gamma_0}{2}\left(\sigma^2+\frac{\hbar}{2}\right)\epsilon_2^{2n} \right\}, \quad (n \gg 1), \tag{9.94}$$

$$\gamma(n) = (\epsilon_1\epsilon_2)^n \exp\left\{ i\gamma_0\sqrt{2\pi\hbar}/k \right\}.$$

In the classical limit we have from (9.94)

$$< \mathbf{G}_n >_{cl}\sim \gamma(n) \exp\left\{ -\frac{\gamma_0}{2}\sigma^2\epsilon_2^{2n} \right\}. \tag{9.95}$$

From (9.94) and (9.95), we derive an equation for the time-scale $n_\hbar$

$$| < \mathbf{G}_n > - < \mathbf{G}_n >_{cl} | = (\epsilon_1\epsilon_2)^n \exp\left\{ -\frac{\gamma_0^2}{2}\epsilon_2^{2n}\sigma^2 \right\} \tag{9.96}$$

$$\times \left\{ 1 - \exp\left[-\frac{\gamma_0^2}{4}\hbar\epsilon_2^{2n} \right] \right\} |_{n=n_\hbar} = \delta \ll 1.$$

Let us introduce the dimensionless parameters,

$$\varepsilon_0 = \sigma^2/\hbar, \quad \hbar_0 = \gamma_0^2\hbar/4. \tag{9.97}$$

Then, the equation (9.96) takes the form

$$e^{-2\hbar_0 \varepsilon_0 \epsilon_2^{2n}} \left(1 - e^{-\hbar_0 \epsilon_2^{2n}}\right) = \delta. \tag{9.98}$$

First, consider the condition

$$\varepsilon_0 \ll 1. \tag{9.99}$$

The condition (9.99) means that the initial packet is very narrow. Then, we find from (9.98) an estimate for the time-scale $n_\hbar$

$$n_\hbar \sim \frac{1}{2 \ln \epsilon_2} \ln \frac{\delta}{\hbar}. \tag{9.100}$$

The estimate for $n_\hbar$ will be different from (9.100) when ε_0 is rather big, and the following condition is satisfied

$$2\hbar_0 \varepsilon_0 \epsilon_2^{2n} \sim 1. \tag{9.101}$$

From (9.101) we find the characteristic crossover-time n_c

$$n_c \sim \frac{1}{2 \ln \epsilon_2} \ln \frac{1}{2\hbar_0 \varepsilon_0}. \tag{9.102}$$

By setting $n_\hbar \sim n_c$ we find a relation between parameters $\hbar$ and σ, when the dependence $n_\hbar$ has a crossover. A rough estimate arises from (9.100) and (9.102)

$$\varepsilon_0 = \sigma^2/\hbar \sim 25. \tag{9.103}$$

A numerical experiment in the situation of weak chaos (9.82) is made in the following way. As discussed in section 9.7a, in this case the quantum map (9.27) gives the time-scale $n_\hbar \sim 1/\hbar$, when quantum dynamics begins to differ significantly from classical. As we are interested here in the logarithmically small time-scale $n_\hbar \sim \ln(1/\hbar)$, in order to calculate the time-scale $n_\hbar$ we used in the numerical experiment the classical map (9.12). All quantum effects which we studied are connected with the initial condition (9.89) which includes quantum parameter (Planck constant) and a classical parameter (the width of initial packet), σ^2. Fig. 12 shows the dependence $n_\hbar$ for the case of narrow initial packet ($\varepsilon_0 \ll 1$). As expected, in this case the dependence of $n_\hbar$ closely follows (9.100). Numerical calculations show (see Figs. 13), that the crossover in the dependence $n_\hbar$ appears in the region of relatively large values of ε_0: $\varepsilon_0 \sim 5$, which is in satisfactory agreement with the rough estimate (9.103). The results shown in Figs. 12,13, are calculated by using the operator $\mathbf{G}(t) = \hbar \mathbf{a}^+(t)\mathbf{a}(t)$.

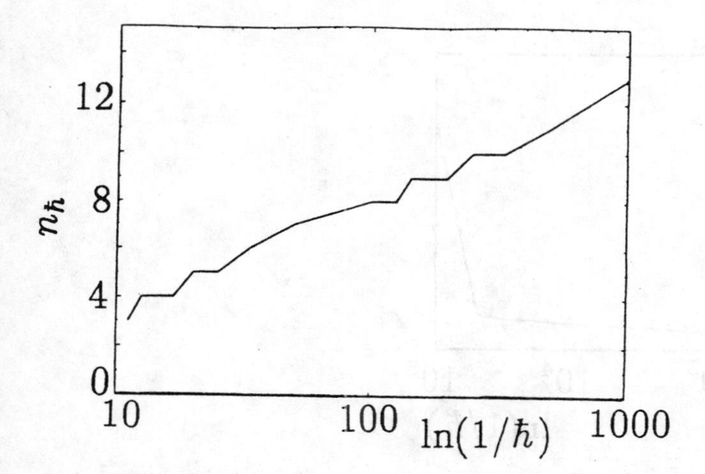

Fig. 12. Dependence $n_\hbar$ in the case of a narrow initial wave packet; $\sigma^2 = 10^{-6}$. The number of trajectories used $M = 10^4; \delta = 2 \cdot 10^{-2}$. Normalization of σ^2 and $\hbar$ by $2I_0/\pi^2$ was used.

9.8 Conclusion

Investigation of quantum dynamics in the system (9.1) shows that under conditions of strong chaos in the classical approximation (K_H $gg1$) the quantum effects can play a rather significant role, and exhibit themselves on logarithmically small time-scale (9.42). These effects are especially important when measuring quantum correlation functions.

Our numerical calculations show, in the case of weak chaos and narrow initial wave packet, that the dependence of $\tau_\hbar$ is close to the logarithmic law (9.42). Upon increasing the width of the initial wave packet the dependence of $\tau_\hbar$ has a crossover, and for rather wide initial wave packet ($\varepsilon_0 \gg 1$), the time $n_\hbar$ significantly increases. So, in this case quantum effects are not so important as under conditions of strong chaos, and reveal themselves at considerably larger time-scale then (9.42). The problem of weak quantum chaos needs additional investigation, including numerical calculations of its dynamics for long times.

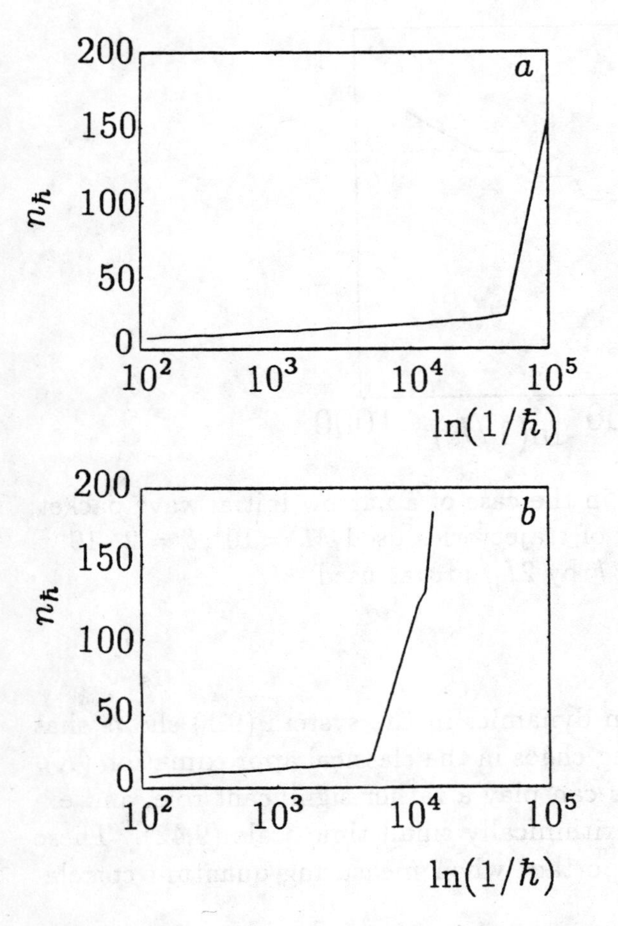

Fig. 13. Dependence of $n_\hbar$ for different values of σ^2. The number of trajectories used $M = 3 \cdot 10^3; \delta = 2 \cdot 10^{-2}$; (a) $\sigma^2 = 10^{-4}$; (b) $\sigma^2 = 9 \cdot 10^{-4}$. Normalization of σ^2 and $\hbar$ by $2I_0/\pi^2$ was used.

10 Characteristic Times for Chaotic Dynamics in Wigner Representation

10.1 Introduction

As was already discussed above, the investigation of quantum chaos in simple models with 1.5 degrees of freedom shows that even in the deep quasiclassical region quantum effects can play an essential role. One of the main phenomena is the existence of the classical-quantum time-scale $\tau_\hbar \sim \ln \kappa, (\kappa \sim 1/\hbar)$ for classically chaotic systems. Another important quantum effect in the region of parameters of quantum chaos is the suppression of quantum diffusion relative to the corresponding classical systems [5,71]. This phenomenon can be important, for example, in the case of strong excitation of quantum systems in the quasiclassical region by external coherent fields (hydrogen atom, molecules, etc. [14,18,53]). The characteristic time-scale τ_D when quantum effects lead to the suppression of classical diffusion is investigated in [5]. As shown in [5], under conditions of strong classical chaos $\tau_D \sim \kappa^2 \gg \tau_\hbar$.

In this section we shall investigate the time-scales $\tau_\hbar$ and τ_D from a viewpoint of the Wigner representation. Stochastic motion in classical systems may be described statistically, and it is natural to use in the quantum analysis of such systems the quantum analog of the classical phase space distribution function. In the Wigner representation the evolution of the quantum system is described by the quantum-mechanical Wigner distribution function, and may be presented in the proper phase space [31].

Actually, the Wigner distribution function is only one of a continuous set of quantum distribution functions of the same character (see also section 9.7b). Namely, there exists, for example, a more general s-parametrized quasidistribution $W(\alpha^*, \alpha, s; \hat{\rho})$ (see section 10.2), where s is a continuous parameter [72]. The Wigner distribution function corresponds to $s = 0$. Some of these quantum quasidistribution functions have disadvantages. For example, they may not be positive definite, and may make no sence for certain initial conditions (see, for example, [73]). Hence, it may be rather difficult to extract physical information from these quasidistribution functions. All these problems vanish if we deal directly with the c-number equations for time-dependent expectation values discussed above. Nevertheless, it is useful to discuss here how the characteristic time-scales $\tau_\hbar$ and τ_D appear in the Wigner representation.

Section 10 is organized in the following way. In section 10.2 we consider s-ordered quantum quasidistribution functions. These functions can be used for describing quantum dynamics of various systems, and are quantum equivalents of the classical Liouville function. In section 10.3 we present the Wigner function in "action-angle" variables. These variables are espesially useful when it is necessary to separate fast and slow dynamics. Such problems occur, for example, when analysing quantum dynamics of classically chaotic systems. In section 10.4 a well known model of the quantum kicked rotator is considered, and the Wigner function for this model is constructed in "action-angle" representation. In section 10.5 the quantum equations of motion ("quantum map") are derived for the quantum kicked rotator. These equations are shown to be completely equivalent to the initial Wigner function in "action-angle" representation. In section 10.6 a "classical model of quantum stochasticity" is analyzed, which approximately describes the quantum kicked rotator in the quasiclassical region of parameters. In this section the characteristic time-scales τ_D and $\tau_\hbar$, which appear in this system, are discussed on the basis of a "classical model of quantum stochasticity". In section 10.7 a discussion of the results is given.

10.2 The S-Ordered Quantum Density Matrix

In this section we shall present a quantum quasidistribution function depending on the continuous parameter s [72]. At the end of this section we shall present a simple example of a quasidistribution function for the Hamiltonian (3.18).

Let us start our considerations with the equation for the density matrix $\hat{\rho}$ for an arbitrary Hamiltonian $\mathbf{H}$

$$i\hbar\dot{\hat{\rho}} = [\mathbf{H}, \hat{\rho}]. \tag{10.1}$$

Introduce a generalized shift operator $\hat{D}(\xi^*, \xi; s)$ given by

$$\hat{D}(\xi^*, \xi; s) = \exp\left(\xi \mathbf{a}^+ - \xi^* \mathbf{a} + \frac{1}{2} s |\xi|^2\right), \tag{10.2}$$

where $\mathbf{a}^+$ and $\mathbf{a}$ are creation and annihilation operators. By using (10.2) and the commutation relation $[\mathbf{a}, \mathbf{a}^+] = 1$ we derive the following formulae

$$\mathbf{a}^+ \hat{D}(\xi^*, \xi; s) = \left[\frac{(1-s)}{2}\xi^* + \frac{\partial}{\partial \xi}\right] \hat{D}(\xi^*, \xi; s), \tag{10.3}$$

$$\hat{D}(\xi^*, \xi; s)\mathbf{a}^+ = \left[-\frac{(1+s)}{2}\xi^* + \frac{\partial}{\partial \xi}\right] \hat{D}(\xi^*, \xi; s),$$

$$\mathbf{a}\hat{D}(\xi^*, \xi; s) = \left[\frac{(1+s)}{2}\xi - \frac{\partial}{\partial \xi^*}\right] \hat{D}(\xi^*, \xi; s),$$

$$\hat{D}(\xi^*, \xi; s)\mathbf{a} = \left[-\frac{(s-1)}{2}\xi - \frac{\partial}{\partial \xi^*}\right] \hat{D}(\xi^*, \xi; s).$$

We write an arbitrary operator $\mathbf{A}(\mathbf{a}^+, \mathbf{a})$ in s-ordered form as

$$\mathbf{A} = \sum_{n,m} A^{(s)}_{n,m} \{(\mathbf{a}^+)^n \mathbf{a}^m\}_s. \tag{10.4}$$

The s-ordering is defined by

$$\{(\mathbf{a}^+)^n \mathbf{a}^m\}_s \equiv \left(\frac{\partial^n}{\partial \xi^n}\right)\left(-\frac{\partial^m}{\partial (\xi^*)^m}\right) \hat{D}(\xi^*, \xi; s)|_{\xi^*=\xi=0}. \tag{10.5}$$

We present some examples,

$$\{\mathbf{a}^+\mathbf{a}\}_0 = \frac{1}{2}(\mathbf{a}^+\mathbf{a} + \mathbf{a}\mathbf{a}^+), \tag{10.6}$$

$$\{\mathbf{a}^+\mathbf{a}^2\}_0 = \frac{1}{3}(\mathbf{a}^+\mathbf{a}^2 + \mathbf{a}\mathbf{a}^+\mathbf{a} + \mathbf{a}^2\mathbf{a}^+).$$

Let us now calculate the average value of the s-ordered operator $\mathbf{A}$ over the density matrix $\hat{\rho}$. We have

$$< \mathbf{A} >_\rho \equiv Tr[\mathbf{A}\hat{\rho}] = \sum_{n,m} A^{(s)}_{n,m} Tr[\{(\mathbf{a}^+)^n \mathbf{a}^m\}_s \hat{\rho}] \tag{10.7}$$

$$= \sum_{n,m} A^{(s)}_{n,m} \left(\frac{\partial^n}{\partial \xi^n}\right)\left(-\frac{\partial^m}{\partial(\xi*)^m}\right) \chi(\xi^*, \xi, s; \hat{\rho})|_{\xi^*=\xi=0},$$

where the following definition is introduced

$$\chi(\xi^*, \xi, s; \hat{\rho}) = Tr[\hat{D}(\xi^*, \xi; s)\hat{\rho}] \equiv \frac{1}{\pi}\int d^2\alpha W(\alpha^*, \alpha, s; \hat{\rho}) e^{\alpha^*\xi - \alpha\xi^*}. \tag{10.8}$$

We shall call the function $W(\alpha^*, \alpha, s; \hat{\rho})$

$$W(\alpha^*, \alpha, s; \hat{\rho}) = \frac{1}{\pi}\int d^2\xi \cdot \chi(\xi^*, \xi, s; \hat{\rho}) e^{\alpha\xi^* - \alpha^*\xi} \tag{10.9}$$

a s-quasidistribution function. Using the s-quasidistribution function we write (10.7) in the form

$$< \mathbf{A} >_\rho = \frac{1}{\pi}\sum_{n,m} A^{(s)}_{n,m} \left(\frac{\partial^n}{\partial \xi^n}\right)\left(-\frac{\partial^m}{\partial(\xi*)^m}\right) \tag{10.10}$$

$$\int d^2\alpha W(\alpha^*, \alpha, s; \hat{\rho}) e^{\alpha^*\xi - \alpha\xi^*}|_{\xi^*=\xi=0}$$

$$= \frac{1}{\pi}\sum_{n,m} A^{(s)}_{n,m} \int d^2\alpha (\alpha^*)^n (\alpha)^m W(\alpha^*, \alpha, s; \hat{\rho})$$

$$= \frac{1}{\pi}\int d^2\alpha A^{(s)}(\alpha^*, \alpha) W(\alpha^*, \alpha, s; \hat{\rho}),$$

where

$$A^{(s)} \equiv \sum_{n,m} A^{(s)}_{n,m} \{(\alpha^*)^n (\alpha)^m\}_s. \tag{10.11}$$

Now we derive an evolution equation for the s-quasidistribution function $W(\alpha^*, \alpha, s; \hat{\rho})$. For this we use the following formula (see (10.8) and (10.9))

$$W(\alpha^*, \alpha, s; \hat{A}) = \frac{1}{\pi}\int d^2\xi e^{\alpha\xi^* - \alpha^*\xi} Tr[\hat{A}\hat{D}(\xi^*, \xi; s)], \tag{10.12}$$

where **A** is an arbitrary operator. Assuming $\hat{A} \equiv \hat{\rho}$ in (10.12), and recalling that the density matrix $\hat{\rho}$ satisfies to the equation (10.1), we have from (10.12)

$$i\hbar \frac{\partial W(\alpha^*, \alpha, s; \hat{\rho})}{\partial t} = \frac{1}{\pi}\int d^2\xi e^{\alpha\xi^* - \alpha^*\xi} \tag{10.13}$$

$$\times \{Tr[D(\xi^*, \xi; s)\mathbf{H}\hat{\rho}] - Tr[\hat{\rho}\mathbf{H}D(\xi^*, \xi; s)]\}.$$

To derive a closed differential equation for the s-quasidistribution function $W(\alpha^*, \alpha, s; \hat{\rho})$ we write an arbitrary Hamiltonian $\mathbf{H}$ in the normal ordering form

$$\mathbf{H} = \sum_{n,m} H^{(N)}_{n,m}(\mathbf{a}^+)^n \mathbf{a}^m. \tag{10.14}$$

Using (10.14) we have from (10.13)

$$i\hbar \frac{\partial W(\alpha^*, \alpha, s; \hat{\rho})}{\partial t} = \sum_{n,m} H^{(N)}_{n,m} \frac{1}{\pi} \int d^2\xi e^{\alpha\xi^* - \alpha^*\xi} \tag{10.15}$$

$$\times \left\{Tr[\hat{D}(\xi^*, \xi; s)(\mathbf{a}^+)^n \mathbf{a}^m \hat{\rho}] - Tr[\hat{\rho}(\mathbf{a}^+)^n \mathbf{a}^m \hat{D}(\xi^*, \xi; s)]\right\}.$$

Then, using (10.2) and (10.3) implies the following relations

$$\hat{D}(\xi^*, \xi; s)(\mathbf{a}^+)^n \mathbf{a}^m = \left[-\frac{(1+s)}{2}\xi^* + \frac{\partial}{\partial \xi}\right]^n \left[\frac{(s-1)}{2}\xi - \frac{\partial}{\partial \xi^*}\right]^m$$

$$(10.16) \times \hat{D}(\xi^*, \xi; s),$$

$$(\mathbf{a}^+)^n \mathbf{a}^m \hat{D}(\xi^*, \xi; s) = \left[\frac{(1+s)}{2}\xi - \frac{\partial}{\partial \xi^*}\right]^m \left[\frac{(1-s)}{2}\xi^* + \frac{\partial}{\partial \xi}\right]^n$$

$$\times \hat{D}(\xi^*, \xi; s).$$

Expressions (10.16) allow one to express equation (10.15) as follows

$$i\hbar \frac{\partial W(\alpha^*, \alpha, s; \hat{\rho})}{\partial t} = \sum_{n,m} H^{(N)}_{n,m} \cdot \frac{1}{\pi} \int d^2\xi e^{\alpha\xi^* - \alpha^*\xi} \tag{10.17}$$

$$\times \{\left[-\frac{(1+s)}{2}\xi^* + \frac{\partial}{\partial \xi}\right]^n \left[\frac{(s-1)}{2}\xi - \frac{\partial}{\partial \xi^*}\right]^m$$

$$- \left[\frac{(1+s)}{2}\xi - \frac{\partial}{\partial \xi^*}\right]^m \left[\frac{(1-s)}{2}\xi^* + \frac{\partial}{\partial \xi}\right]^n \}\chi(\xi^*, \xi, s; \hat{\rho}).$$

In (10.17) the definition (10.8) is used. Now we use the formulae which follow directly from (10.8) and (10.9)

$$\frac{1}{\pi} \int d^2\xi e^{\alpha\xi^* - \alpha^*\xi} \left[\frac{(s-1)}{2}\xi - \frac{\partial}{\partial \xi^*}\right] \chi(\xi^*, \xi, s; \hat{\rho}) \tag{10.18}$$

$$= \left[\alpha - \frac{(s-1)}{2}\frac{\partial}{\partial\alpha^*}\right] W(\alpha^*, \alpha, s; \hat{\rho}),$$

$$\frac{1}{\pi}\int d^2\xi e^{\alpha\xi^* - \alpha^*\xi}\left[-\frac{(1+s)}{2}\xi^* + \frac{\partial}{\partial\xi}\right]\chi(\xi^*, \xi, s; \hat{\rho})$$

$$= \left[\alpha^* - \frac{(s+1)}{2}\frac{\partial}{\partial\alpha}\right] W(\alpha^*, \alpha, s; \hat{\rho}),$$

$$\frac{1}{\pi}\int d^2\xi e^{\alpha\xi^* - \alpha^*\xi}\left[\frac{(1+s)}{2}\xi^* - \frac{\partial}{\partial\xi^*}\right]\chi(\xi^*, \xi, s; \hat{\rho})$$

$$= \left[\alpha - \frac{(s+1)}{2}\frac{\partial}{\partial\alpha^*}\right] W(\alpha^*, \alpha, s; \hat{\rho}),$$

$$\frac{1}{\pi}\int d^2\xi e^{\alpha\xi^* - \alpha^*\xi}\left[-\frac{(s-1)}{2}\xi^* - \frac{\partial}{\partial\xi}\right]\chi(\xi^*, \xi, s; \hat{\rho})$$

$$= \left[\alpha^* - \frac{(s-1)}{2}\frac{\partial}{\partial\alpha}\right] W(\alpha^*, \alpha, s; \hat{\rho}).$$

Finally we derive from (10.17), (10.18) an equation for the evolution of the s-quasidistribution function $W(\alpha^*, \alpha, s; \hat{\rho})$

$$i\hbar\frac{\partial W(\alpha^*, \alpha, s; \hat{\rho})}{\partial t} = \sum_{n,m} H_{n,m}^{(N)}\{\left[\alpha^* - \frac{(s+1)}{2}\frac{\partial}{\partial\alpha}\right]^n\left[\alpha - \frac{(s-1)}{2}\frac{\partial}{\partial\alpha^*}\right]^m$$

$$-\left[\alpha - \frac{(s+1)}{2}\frac{\partial}{\partial\alpha^*}\right]^m\left[\alpha^* - \frac{(s-1)}{2}\frac{\partial}{\partial\alpha}\right]^n\}W(\alpha^*, \alpha, s; \hat{\rho}). \quad (10.19)$$

As an example, consider the equation for the s-quasidistribution function $W(\alpha^*, \alpha, s; \hat{\rho})$ with Hamiltonian (3.18) and for $s = 1$. In this case $m, n \leq 2$ and (cf. the equation after (4.34))

$$i\frac{\partial W}{\partial t} = \{\frac{\partial}{\partial\alpha^*}[(\omega + \mu\hbar)\alpha^* + 2\mu\hbar\alpha\alpha^{*2}] \quad (10.20)$$

$$-\frac{\partial}{\partial\alpha}[(\omega + \mu\hbar)\alpha + 2\mu\hbar\alpha^*\alpha^2]\}W + \mu\hbar\left(\frac{\partial^2}{\partial\alpha^2}\alpha^2 - \frac{\partial^2}{\partial\alpha^{*2}}\alpha^{*2}\right)W.$$

If we choose the parameter $s = 0$, the corresponding equation (10.20) will include fourth order derivatives. In general case the following "heat-conduction" equation for the s-quasidistribution function $W(\alpha^*, \alpha, s; \hat{\rho})$ arises

$$\frac{\partial W(\alpha^*, \alpha, s; \hat{\rho})}{\partial s} = \frac{1}{2\pi} \int d^2\xi e^{\alpha\xi^* - \alpha^*\xi} |\xi|^2 \chi(\xi^*, \xi, s; \hat{\rho}) \tag{10.21}$$

$$= -\frac{1}{2} \frac{\partial^2}{\partial\alpha\partial\alpha^*} W(\alpha^*, \alpha, s; \hat{\rho}) = -\frac{1}{8} \left(\frac{\partial^2}{\partial x^2} + \frac{\partial^2}{\partial y^2} \right) W(\alpha^*, \alpha, s; \hat{\rho}),$$

where we have put $\alpha = x + iy$. In this "heat-conduction" equation s is an effective time, and W is an effective temperature. The considerations presented above show that the behavior of the quasidistribution function $W(\alpha^*, \alpha, s; \hat{\rho})$ significantly depends on the type of ordering made (the value of the parameter s). In spite of this apparent nonuniqueness, the quantum-dynamical expectation value (10.7) does not depend on s. This circumstance demonstrates the power and convenience of the c-number equations for quantum expectation values discussed in section 3.

10.3 Wigner Representation in "Action-Angle" Variables

As a rule the process of the evolution of a quantum system in the Wigner representation is described in "coordinate-momentum" variables (x, p) [31,73-76]. However, in the classical case there are some advantages in describing the dynamics of the nonlinear systems in "action-angle" variables ensuring the separation of the motion into the fast (phase) and slow (e.g. diffusion in action). Therefore in the quantum analysis of such systems in Wigner representation it is useful to generalize it for the case of the "action-angle" variables. In the case of systems whose spectrum is unrestricted on both sides ("rotational" spectrum) this modification was proposed in [77] and rigorously proved on the basis of group-theoretical analysis in [78].

Below we introduce the Wigner representation for the case of "action-angle" variables on the basis of the quantum-mechanical "action-angle" representation proposed in [6]. The representation introduced in [6] is based on the following. Let the Hamiltonian of the system in the initial representation have the form

$$\mathbf{H} = \mathbf{H}_0 + \mathbf{H}_{int}, \tag{10.22}$$

and suppose the spectrum of $\mathbf{H}_0$ is discrete. Instead of the wave function in the initial representation a new wave function $\psi(\theta, t)$ periodic in phase θ is introduced: $\psi(\theta+2\pi, t) = \psi(\theta, t)$. This periodic function satisfies the Schrödinger equation

$$i\hbar \frac{\partial \psi(\theta, t)}{\partial t} = \mathbf{H}(\theta, \hat{n}, t)\psi(\theta, t), \quad \hat{n} \equiv -i\frac{\partial}{\partial \theta} \tag{10.23}$$

with Hamiltonian

$$\mathbf{H}(\theta, \hat{n}, t) = \mathbf{H}_0(\hat{n}) + \sum_{k=1}^{\infty} \left[e^{-ik\theta} \mathbf{G}_k(\hat{n}, t) + \mathbf{G}_k^{+}(\hat{n}, t) e^{ik\theta} \right], \tag{10.24}$$

where the operator functions $\mathbf{H}_0$ and $\mathbf{G}_k(\hat{n}, t)$ are unambiguously determined by the matrix elements of the unperturbed Hamiltonian, and the interaction Hamiltonian is calculated in the basis of eigenfunctions of unperturbed Hamiltonian. The Fourier coefficients of the expansion $c_n(t)$ of the wave function $\psi(\theta, t)$

$$\psi(\theta, t) = \sum_n c_n(t) e^{in\theta} \tag{10.25}$$

determine the probability amplitudes of the system being populated at the n-th level of the unperturbed Hamiltonian $\mathbf{H}_0$. The summation in (10.25) depends on the limiting character of the spectrum of the operator $\mathbf{H}_0$. For example, when the spectrum is bounded below the index n in (10.25) takes on the values $n = 0, 1, 2, \ldots$. The initial Cauchy problem for (10.23)-(10.25) is determined on the subspace of the functions $\{e^{in\theta}; n \geq 0\}$ which is invariant relative to the action of the evolution operator $\hat{U}(t, t_0)$ with the Hamiltonian $\mathbf{H}(\theta, \hat{n}, t)$ [6]. Following [6] we call the representation (10.23)-(10.25) an "action-angle" representation for as $\hbar \to 0$ equation (10.23) transforms into the Hamiltonian-Jacobi equation in "action-angle" variables.

The arbitrary operator $\mathbf{A}$ of the initial representation may be represented in the "action-angle" representation as

$$\mathbf{A} \equiv \mathbf{A}(\theta, \hat{n}) = \mathbf{A}(\theta + 2\pi, \hat{n}). \tag{10.26}$$

Define the Weyl transformation of the operator $\mathbf{A}(\theta, \hat{n})$ (10.26) in the c-number function $a(\varphi, p)$ in the following way

$$a(\varphi, p) = Tr\left[\mathbf{A}(\theta, \hat{n})\hat{\Delta}(\varphi, p)\right]; \tag{10.27}$$

$$0 \leq \varphi < 2\pi, \quad p = 0, \pm 1, ...,$$

where the symbol Tr denotes the trace of the operator, and

$$\hat{\Delta}(\varphi, p) = \frac{1}{2\pi} \sum_{m=-\infty}^{\infty} \int_{-\pi}^{\pi} d\xi \exp[i(\varphi - \theta)m + i(p - \hat{n})\xi]. \tag{10.28}$$

Then, the inverse transformation has the form

$$\mathbf{A}(\theta, \hat{n}) = \frac{1}{2\pi} \sum_{m=-\infty}^{\infty} \int_{-\pi}^{\pi} d\varphi \cdot a(\varphi, p)\hat{\Delta}(\varphi, p). \tag{10.29}$$

Since the "action-angle" representation [6] does not use explicitly the angle operator θ, but only the phase operators $\Lambda^{\pm}(\theta) = \exp(\pm i\theta)$, a formal writing of the exponent in (10.28) implies

$$e^{-im\theta - i\xi\hat{n}} = e^{i(m\xi/2)}e^{-im\theta}e^{-i\xi\hat{n}} == e^{-i(m\xi/2)}e^{-i\xi\hat{n}}e^{-im\theta}. \tag{10.30}$$

Formula (10.29) is derived analogously to the case of the (x, p)-Wigner representation (see Appendix C). Furthermore, we call the function $a(\varphi, p)$ the image of the operator $\mathbf{A}(\theta, \hat{n})$.

We give, in conclusion of this section, the formula for the convolution of the product of two operators which we shall need later,

$$Tr[\mathbf{A}(\theta, \hat{n})\mathbf{B}(\theta, \hat{n})] = \frac{1}{2\pi} \sum_{p=-\infty}^{\infty} \int_{0}^{2\pi} d\varphi a(\varphi, p)b(\varphi, p), \tag{10.31}$$

where $a(\varphi, p)$ and $b(\varphi, p)$ are the images of the operators $\mathbf{A}$ and $\mathbf{B}$, respectively (see Appendix C).

10.4 Evolution of the Quantum Kicked Rotator in the Wigner Representation

We consider the well known quantum kicked rotator [5] in the Wigner representation. Various physical problems reduced to this model [46,48]. Solutions of this system yield the form of the quasi-energy spectrum, the time behaviour of quantum correlation functions, the connection with Anderson localization in solid state systems, and time-scales $\tau_\hbar$ and τ_D [5,6,12,49,53,79-84].

In the classical case the Hamiltonian of the kicked rotator has the form

$$H_{cl}(\varphi, I, t) = \frac{\mu I^2}{2} + \varepsilon f(\varphi) \sum_{n=-\infty}^{\infty} \delta(t - nT), \tag{10.32}$$

where $f(\varphi + 2\pi) = f(\varphi)$, and ε and μ are the parameters of perturbation and nonlinearity, respectively. The classical map according to (10.32) has the form

$$I_{t+1} = I_t - \varepsilon \frac{df(\varphi)}{d\varphi}|_{\varphi=\varphi_t}, \tag{10.33}$$

$$\varphi_{t+1} = \varphi_t + \mu T I_{t+1},$$

where the following definitions are used

$$I_t \equiv I(nT - 0), \quad \varphi_t \equiv \varphi(nT - 0). \tag{10.34}$$

The equation for the classical distribution function is of the form according to (10.33)

$$\rho_{t+1}(\varphi, I) = \int_0^{2\pi} d\varphi' \int_{-\infty}^{\infty} dI' \mathbf{G}_{cl}(\varphi, I|\varphi', I')\rho_t(\varphi', I'), \tag{10.35}$$

where $\mathbf{G}_{cl}$ is the classical Green function

$$\mathbf{G}_{cl}(\varphi, I|\varphi', I') = \tilde{\delta}(\varphi' + \mu T I - \varphi)\delta\left(I' - \varepsilon\frac{df}{d\varphi}|_{\varphi=\varphi'} - I\right). \tag{10.36}$$

Henceforth $\tilde{\delta}$ denotes a periodic δ-function with period 2π. The operator (10.36) preserves the area bounded by a closed contour on the plane (φ, I). However, when increasing t the initial contour distorts and gets a rather complicated structure. This circumstance leads to fast decay of the phase correlation functions, thereby enabling us to describe the system (10.32) in a statistical approach [57].

We describe the evolution of the kicked rotator in the quantum case by the Wigner function, which is the Weyl transformation of the density matrix operator

$$\hat{\rho}(t) = |\psi(t)><\psi(t)|, \tag{10.37}$$

where the wave function $|\psi(t)>\equiv \psi(\theta, t)$ satisfies the Schrödinger equation

$$i\hbar \frac{\partial \psi(\theta, t)}{\partial t} = \mathbf{H}\psi(\theta, t), \tag{10.38}$$

$$\mathbf{H} = -\frac{\mu \hbar^2}{2}\frac{\partial^2}{\partial \theta^2} + \varepsilon f(\theta) \sum_{n=-\infty}^{\infty} \delta(t - nT).$$

From (10.37) and (10.38) we have

$$\hat{\rho}_{t+1} = \hat{U}\hat{\rho}_t \hat{U}^+, \tag{10.39}$$

$$\hat{U} = \exp\left(i\pi\zeta \frac{\partial^2}{\partial \theta^2}\right) \exp\left[-i\kappa f(\theta)\right],$$

where $\hat{U}$ is the evolution operator; and ζ and κ are parameters

$$\zeta = \frac{\mu \hbar T}{2\pi}, \quad \kappa = \frac{\varepsilon}{\hbar}. \tag{10.40}$$

In (10.39) t is a discrete time: $t = 0, 1, ...$; $\hat{\rho}_t \equiv \hat{\rho}(nT - 0)$.

The model (10.38) itself contains the most interesting peculiarities of dynamical behavior of quantum systems that are chaotic in the classical limit. Such a model can approximately describe a real physical system with a non-equidistant unperturbed spectrum influenced by a strong external field which has a large number of harmonics. The system (10.38) with $f(\theta) = \cos\theta$ was numerically investigated for the first time in [71]. The main result derived in [71] is the discovery of considerable differences in the behavior of a quantum system in comparison with a classical one in the case when the classical motion is chaotic with a characteristic linear increase of the mean energy E_t of the system. In particular, the growth of the energy corresponds to the classical only during some finite time $t < t_*$, which depends essentially on the parameters of the system; then it sharply drops. Further investigations of the peculiarities of such behavior is made in [5,85], where it is shown analytically that for a rational value of the parameter ζ in (10.40) (the so-called "quantum resonance"), the asymptotic form E_t is quadratic: $E_t \sim t^2$ for $t \to \infty$. In the notation (10.40) the value $\zeta = 2$ corresponds to the so-called main quantum resonance.

The results of numerical investigations of the increase of the averaged energy for the quantum kicked rotator can be formulated in the following way. When the value of the parameter ζ is equal to a "good" rational number $\zeta = r/q$ (r and q are not too large), the increase of the mean energy has a strongly quadratic dependence starting from $t = 0$. When the rational number ζ starts getting worse ($r, q \to \infty$, $r/q \to$ *const*; this corresponds to the transition to an irrational value ζ), the law of the increase of the mean energy shows the characteristic limitation of classical diffusion for relatively large time, and only after this time does asymptotic quadratic behaviour start to manifest itself (see, also [86]). It is interesting to note that numerical experiments always have the case of quantum resonance in some sense, because of the finite digits of all numbers. However, the behaviour of the system should correspond to the nonresonant case during enomously large times. It can be shown that the wave functions for two close values of parameter ζ differing by a small value $\sim \delta\zeta = |\zeta_1 - \zeta_2|$ is given by

$$\frac{1}{2\pi}\int_0^{2\pi} d\theta |\psi_{\zeta_1}(\theta, t) - \psi_{\zeta_2}(\theta, t)|^2 < \kappa^2 t^3 \delta\zeta,$$

where κ is introduced in (10.41). Therefore, for our purpose we can confine ourself to the analysis of the system (10.38) only for rational values of ζ. The case of rational values of ζ is also of special interest since it corresponds to resonances

$$\omega_n \equiv (E_{n+1} - E_n)/\hbar = (2n+1)r\Omega/q, \quad (E_n = \mu\hbar^2 n^2/2, \Omega = 2\pi/T)$$

between harmonics of the frequency of the unperturbed system and the harmonics of the frequency of the external field. The above-mentioned peculiarities of dynamics of a quantum kicked rotator are investigated from different points of view in, for example, [6,86-89]. Below, in this section we analyze the system (10.38) using the Wigner function formalizm.

The results of section 10.3 lead analogously to the case of (x,p)-representation [86,90] to an equation of evolution for the Wigner function (see Appendix C)

$$\rho_{t+1}(\varphi,p)=\int_0^{2\pi}d\varphi'\sum_{p'=-\infty}^{\infty}\mathbf{G}_q(\varphi,p|\varphi',p')\rho_t(\varphi',p'), \tag{10.42}$$

where the quantum Green function is given by

$$\mathbf{G}_q(\varphi,p|\varphi',p')=\frac{1}{(2\pi)^2}\sum_{m=-\infty}^{\infty}\int_{-\pi}^{\pi}d\xi\cdot\exp\{im(\varphi'-\varphi+2\pi\zeta p')+i\xi(p-p')$$
$$+i\kappa\left[f\left(\varphi'-\frac{2\pi m\zeta+\xi}{2}\right)-f\left(\varphi'+\frac{2\pi m\zeta+\xi}{2}\right)\right]\}. \tag{10.43}$$

The formulae (10.42) and (10.43) are easily shown to transform under $\hbar\to 0$, $p\hbar=I$ into their classical analogs (10.35) and (10.36).

To simplify the analytical expressions we restrict further to the case

$$f(\varphi)=\cos 2\varphi,\quad (f(\theta)=\cos 2\theta)\,. \tag{10.44}$$

Then, from (10.43) we get for the quantum Green function

$$\mathbf{G}_q(\varphi,p|\varphi',p')=\tilde{\delta}(\varphi'+2\pi\zeta p-\varphi)J_{p-p'}(2\kappa\sin 2\varphi'), \tag{10.45}$$

where $J_p(z)$ is the Bessel function.

Using (10.27) and (10.37) we present the Wigner function $\rho(\varphi,p)$ in the familiar form

$$\rho(\varphi,p)=\frac{1}{2\pi}\int_{-\pi}^{\pi}d\xi\cdot e^{ip\xi}\psi^*\left(\varphi+\frac{\xi}{2}\right)\psi\left(\varphi-\frac{\xi}{2}\right). \tag{10.46}$$

Note that in the quantum case, as follows from (10.45), the map for one step of the transformation for φ and p variables acts locally in phase and nonlocally in action. Such a situation is typical for systems kicked by δ-pulses [75,76].

10.5 Quantum Equations of Motion

In the classical case the equations of motion connecting the action I and the phase φ over a single kick for $f(\varphi) = \cos 2\varphi$ have the form (see 10.33)

$$I_{t+1} + I_t + 2\varepsilon \sin 2\varphi_t, \tag{10.47}$$

$$\varphi_{t+1} = \varphi_t + \mu T I_{t+1},$$

with the phase and action changing continuously.

We introduce formally the following transformations

$$p_{t+1} = p_t + [2\kappa \sin 2\varphi_t]_{int} + \Delta p_t, \tag{10.48}$$

$$\varphi_{t+1} = \varphi_t + \mu T \hbar p_{t+1},$$

where $[\cdots]_{int}$ denotes an integer part; p_t changes discretely ($p = 0, \pm 1, ...$); and Δp_t is a quasirandom function taking discrete values and distributed by the law $W(\Delta p_t)$. It is easily seen that when we choose the function $W(\Delta p_t)$ in the form

$$W(\Delta p_t) \tag{10.49}$$

$$= \frac{1}{2\pi} \int_{-\pi}^{\pi} d\xi \cdot \exp\left\{-i\xi\Delta p_t - i\xi\left[2\kappa \sin 2\varphi_t\right]_{int} + i2\kappa \sin 2\varphi_t \sin \xi\right\},$$

equations (10.48) lead to a law of the evolution of the distribution function coinciding with the equation for the evolution of the Wigner function $\rho_t(\varphi, p)$ for the quantum rotator (10.42), (10.45)

$$\rho_t(\varphi, p) = \sum_{p'=-\infty}^{\infty} \int_0^{2\pi} d\varphi \cdot \tilde{\delta}(\varphi' + 2\pi\zeta p - \varphi) J_{p-p'}(2\kappa \sin 2\varphi') \rho_{(t-1}(\varphi', p'). \tag{10.50}$$

In fact, using the transformations (10.48) the equation for the distribution function $D_t(\varphi, p)$ in the general case can be written in the form

$$D_{t+1}(\varphi, p) = \sum_{\Delta p'=-\infty}^{\infty} W(\Delta p') \tag{10.51}$$

$$\times \left[\sum_{p'=-\infty}^{\infty} \int_0^{2\pi} d\varphi' \delta_{p'+[2\kappa \sin 2\varphi']_{int}+\Delta p', p} \cdot \tilde{\delta}(\varphi' + \mu T \hbar p - \varphi) D_t(\varphi', p')\right],$$

which, with the use of (10.49), leads to the equation (10.50), where $D_t(\varphi, p) \equiv \rho_t(\varphi, p)$.

Thus, the formal equations of motion lead to the correct expression for the evolution of the quasidistribution function $\rho_t(\varphi, p)$. Consequently, the evolution of the initial particle distribution on the "quantum phase space" (φ, p) may be, analogously to the classical case, determined on the basis of the equations of motion with the difference that the function $W(\Delta p_t)$ can take on negative values, and that additional specific conditions must be imposed on the initial distribution function $\rho_0(\varphi, p)$ that guarantee its connection with a pure quantum state [73].

The equations of motion (10.48) may be interpreted as the equations of motion for the quantum rotator, since under the condition $\hbar \to 0$ they transform into the corresponding classical equations of motion. Actually, one can easily derive the following expressions for the moments of the function $W(\Delta p)$

$$\sum_{\Delta p=-\infty}^{\infty} W(\Delta p) = 1, \quad \left| \sum_{\Delta p=-\infty}^{\infty} \Delta p W(\Delta p) \right| \leq 1, \tag{10.52}$$

$$\left| \sum_{\Delta p=-\infty}^{\infty} (\Delta p)^2 W(\Delta p) \right| \leq 1,$$

which are valid for arbitrary time t. Put $\hbar p_t = I_t$, multiply the first equation in (10.48) by $\hbar W(\Delta p_t)$, and sum it over Δp_t. Allowing for (10.52) we derive from (10.48) the classical equations of motion (10.47).

To avoid misunderstanding, it is necessary to note that the transformation (10.48) represents another (but absolutely equivalent) way of describing of the dynamics of the system, which is convenient for comparison with the classical case. Transformations (10.48) differ from the classical transformations by the operation of an integer part, and by the discrete quasi-random term Δp_t in the first equation. The first circumstance is due to the fact that the phase space is discrete in action in the (φ, I)-Wigner representation (unlike the (x, p)-Wigner representation). The second circumstance reflects the fact of nonlocality of the quantum map in the action variable.

To analyse the dynamics of the Wigner function (10.50) we choose as the initial condition a state of the rotator with a definite value p

$$\rho_0(\varphi, p) = \frac{1}{2\pi}\delta_{p,p_0}. \tag{10.53}$$

An important circumstance follows immediatly from (10.50). Due to the discreteness of phase space in action $I = \hbar p$ one can choose invariant countable sets

$$\varphi_n = \{\varphi_0 + 2\pi\zeta n\}_{2\pi}, \quad n = 0, \pm 1, ..., \tag{10.54}$$

where $\{\cdots\}_{2\pi}$ means a fractional part on modulo 2π, and φ_0 is an arbitrary initial phase (see Fig. 14). The evolution of the Wigner function $\rho_t(\varphi, p)$ on a set $\{\varphi_n\}$ (10.54) does not depend on the values $\rho_t(\varphi, p)$ in different parts of phase space of measure 1. In the case when ζ is rational, $\zeta = r/q$, the sequence of phases $\{\varphi_n\}$ in (10.54) is finite ($0 \leq n < q$). In the case when ζ is irrational the sequence $\{\varphi_n\}$ is infinite and densely covers an interval $0 \leq \varphi < 2\pi$. This difference leads to a difference in expressions for quantum-mechanical expectation values for an arbitrary operator $< \mathbf{A}(t) >= Tr[\mathbf{A}\hat{\rho}_t]$. In particular, for $\zeta = r/q$ from (10.31) we have

$$< \mathbf{A}(t) >= \int_0^{2\pi/q} d\varphi_0 \left[\sum_{p=-\infty}^{\infty} \sum_{n=0}^{q-1} a(\varphi_n, p)\rho_t^{(\varphi_0)}(\varphi_n, p)\right]. \tag{10.55}$$

Here $a(\varphi_n, p)$ is an image of the operator $\mathbf{A}$; the upper index (φ_0) indicates an explicit dependence of the Wigner function on the phase φ_0.

For our further purposes it is convenient to pass from equation (10.50) to the dynamic equations of motion for p and φ (10.48). For simplicity we confine ourself to the quasiclassical case of $\kappa \gg 1$. Then we have from (10.49)

$$W(\Delta p_t) \approx \frac{1}{2\pi}\int_{-\pi}^{\pi} d\xi \exp[-i\xi\Delta p_t - 2i\kappa(\xi - \sin\xi)\cdot\sin 2\varphi_t] \equiv \tilde{W}(\Delta p_t). \tag{10.56}$$

The function $\tilde{W}(\Delta p_t)$ has the following properties

$$\sum_{\delta p} \tilde{W}(\Delta p) = 1, \quad \sum_{\delta p} \Delta p \tilde{W}(\Delta p) = 0, \quad \sum_{\delta p} (\Delta p)^2 \tilde{W}(\Delta p) = 0. \tag{10.57}$$

These expressions allow to assume that when $\kappa \gg 1$ the contribution of the term Δp_t (see (10.48)) in the diffusion of the action can be considered to be small in comparison with the effects connected with the descreteness of the phase space. The behavior of the rotator in this case can be approximately described by the following map

$$p_{t+1} = p_t + [2\kappa \sin 2\varphi_t]_{int}, \tag{10.58}$$

$$\varphi_{t+1} = \{\varphi_t + 2\pi\zeta p_{t+1}\}_{2\pi}.$$

This model is introduced phenomenologically in [5] to account for the influence of the discreteness of phase space on the behavior of the system. Numerical analysis [5] shows that the model (10.58) is capable of describing the limitation of quantum diffusion qualitatively. Further, following [5], we shall refer to (10.58) as the "classical model of quantum stochasticity".

To estimate the influence of nonlocality in (10.48) we write the kernel of the transformation in (10.45) as the sum of two terms, $\mathbf{G}_q = \mathbf{G}_{cq} + \mathbf{G}_v$. Here

$$\mathbf{G}_{cq}(\varphi, p|\varphi', p') = \tilde{\delta}(\varphi' + 2\pi\zeta p - \varphi) \cdot \delta_{p,p'+[2\kappa \sin 2\varphi']_{int}}, \tag{10.59}$$

provides the Wigner function evolution according to the "classical model of quantum stochasticity" (10.58), and $\mathbf{G}_v = \mathbf{G}_q - \mathbf{G}_{cq}$ is responsible for the nonlocality of the Wigner function (10.45). Then the time-evolution of the Wigner function can be written as

$$\rho_t = (\mathbf{G}_{cq} + \mathbf{G}_v)^t \rho_0, \quad \rho_0 = \frac{1}{2\pi}\delta_{p,0}, \tag{10.60}$$

where $\mathbf{G}_{cq,v}\rho$ has the meaning

$$\mathbf{G}_{cq,v}\rho \equiv \int_0^{2\pi} d\varphi' \sum_{p'} \mathbf{G}_{cq,v}(\varphi, p|\varphi', p')\rho(\varphi', p'). \tag{10.61}$$

The product in (10.60) has 2^t terms which are the product of operators $\mathbf{G}_{cq}$ and $\mathbf{G}_v$ in different combinations. Therefore, it is convenient to represent the expression for the averaged energy of the system in the following form

$$E_t = \frac{1}{2\pi}\int d\varphi \sum_p \left[\mathbf{G}_{cq}^t + \mathbf{G}_{cq}^{t-1}\mathbf{G}_v + \mathbf{G}_{cq}^{t-2}(\mathbf{G}_v^2 + \mathbf{G}_v\mathbf{G}_{cq}) + \mathbf{G}_v(\cdots)\right]\rho_0. \tag{10.62}$$

Using (10.57) it can be shown that at $\kappa \gg 1$ the terms which have $\mathbf{G}_v$ on the right side are relatively small. This means that all of the terms in (10.62) are ordered by the value of this contribution to the averaged energy. It should be noted that only the first term in (10.62) gives a strictly positive contribution. The rest of the terms may have arbitrary signs. Numerical calculations show that the relative contribution to the averaged energy of the terms corresponding to different combinations of the operators $\mathbf{G}_{cq}$ and $\mathbf{G}_v$ is in good agreement with the order of their positions in expression (10.62). In particular, the ratio of the contributions of the last and the first terms in (10.62) is of the order of 1% (for $2\kappa = 10.2; \zeta = 1/16; t = 4$).

10.6 Classical Model of Quantum Stochasticity

The results of the previous section show that when $\kappa \gg 1$ the dynamics of the mean energy for the quantum rotator is mainly determined by the properties of the map for $\mathbf{G}_{cq}$. We now pass on to the analysis of the classical model of quantum stochasticity (10.58) [86,90,91]. In what follows we shall investigate the behavior of the system (10.58) at rational values $\zeta = r/q$. The case of irrational ζ will be considered as the transition $r, q \to \infty$. As mentioned in section 10.5, for rational values ζ in (10.58) the trajectory of each particle lies on an invariant set (see Fig. 14) consisting of q points of phase φ. Since all the values of the phase φ are numbered it is convenient to rewrite the map (10.58) in the form

$$p_{t+1} = p_t + \left[2\kappa \sin 2\left(\varphi_0 + 2\pi\frac{r}{q}n_t\right)\right]_{int}, \tag{10.63}$$

$$n_{t+1} = [n_t + p_{t+1}]_q.$$

In (10.63) n_t is a discrete phase ($0 \leq n_t < q$) on the discrete step of transformation; $[x]_q$ is the integer part modulo q. It is easy to see that the map (10.63) is invariant with respect to the shift $n \rightarrow n+q$ and $p \rightarrow p+q$. Thus, the motion of the particle can be considered on a torus consisting of $N = q \times q$ points, and any trajectory can be characterized by two parameters. The first is the period M-the number of steps before locking of the trajectory (since the number of points on a torus is finite, any trajectory for $M \leq N$ steps will be locked). The second parameter is the number of rotations J around the torus along the axis of action p which the trajectory makes before locking. As examples two trajectories locked on a torus with the number of rotations $J = 1$ and $J = 0$ are shown in Fig. 14.

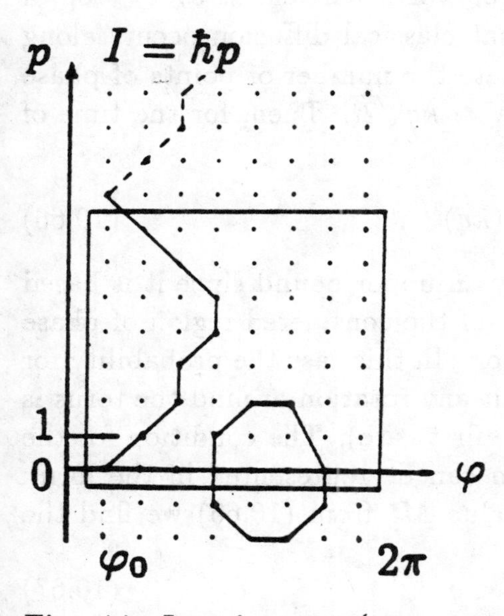

Fig. 14. Invariant set (represented by points) for $\zeta = 1/8$. As an example, two trajectories locked on a torus with the number of rotations equal to 1 and 0 are shown.

Let us discuss the characteristic features of the dynamics of the system (10.63) in different regions of parameters. Two limiting cases are possible: (a) $2\kappa > q$ and (b) $2\kappa \ll q$. In the first case any point of a torus is accessible for a particle at every step. Supposing that the particle moves randomly in the discrete phase space it is easy to estimate the probability that the trajectory will not lock for M steps

$$P(M) \approx \exp(-M^2/2N), \quad (M/N \ll 1). \tag{10.64}$$

Assuming $P(M) \approx 0.5$, we have for a characteristic period of trajectories

$$M^* \approx \sqrt{N}. \tag{10.65}$$

In this case there exists a relative large number of trajectories with a nonzero number of rotations J. Suppose now that we make a transition to the irrational value ζ increasing the values of r and q. Then, the second case occurs: $2\kappa \ll q$. In this case not the whole region of a torus is accessible for a particle at every step of the transformation. Assuming that classical diffusion occurs along the axis of action p we can estimate the number of points of phase space occupied by the particle: $\tilde{N} \approx \kappa q\sqrt{2t}$. Then, for the time of locking $M_1^* \sim \sqrt{\tilde{N}}$ we have

$$M_1^* \sim (\kappa q)^{2/3}. \tag{10.66}$$

The estimate (10.66) is essentially an upper bound since it is based on the assumption of accessibility of the considered region of phase space at each step of transformation. In this case the probability for the trajectory to be locked without any rotation around the torus is increased (in comparison with the first case). The condition for the realization of such a phenomenon can be represented in the form: $\kappa\sqrt{2M_1^*} < q$. Substituting the value M_1^* from (10.66) we find the critical relation

$$2\kappa^2 < q. \tag{10.67}$$

The existence of the critical relation (10.67) for the parameters κ and q is illustrated in Fig. 15a-c, where the trajectories of $q = 101$ particles on a torus with the initial distribution $p_0 = 0$ are presented. The value of the parameter ζ is equal $\zeta = 10/101$. In the case presented in Fig. 15a ($2\kappa = 5.0$) there is no trajectory with the rotations on a torus along the axis of action p. Fig. 15b corresponds to the value $2\kappa = 10.0$. In this case six trajectories have made one rotation. In the case shown in Fig. 15c ($2\kappa = 20.0$) four trajectories have the value $J = \pm 1$, forteen: $J = \pm 2$, two: $J = \pm 3$.

We now pass to the analysis of the quantum diffusion of the mean energy for the system (10.58). Consider the set of particles uniformly distributed on the interval $[0, 2\pi]$. Choose an arbitrary point from the set. Since the term $[2\kappa \sin 2\varphi_t]_{int}$ represents a step function, and locking of the trajectory occurs on a torus after a finite number of steps, we can surround the chosen point by some small (but finite) region $\Delta\varphi_t$, any point from which is moving according to the same law.

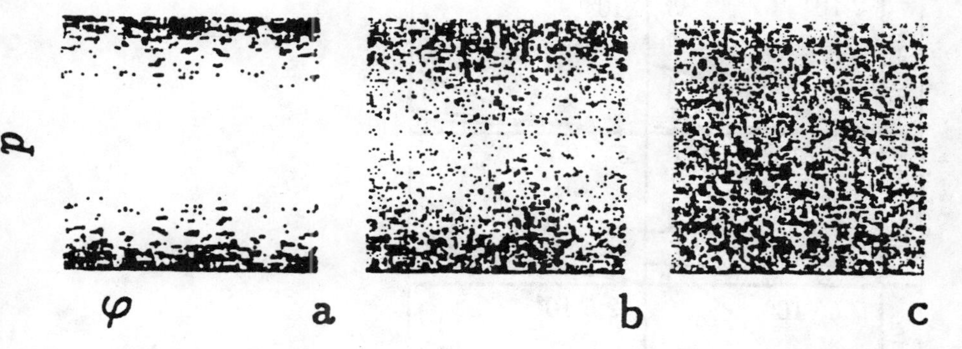

Fig. 15. Numerical data for the map (10.63). The trajectories on a torus of $q = 101$ particles "starting" from the bottom line; $\zeta = 10/101; \varphi_0 = 0$; (a) $2\kappa = 5$; all trajectories have $J = 0$; (b) $2\kappa = 10$; six trajectories have $J = \pm 1$; (c) $2\kappa = 20$; four trajectories have $J = \pm 1$; fourteen $J = \pm 2$; two $J = \pm 3$.

Thus, the time-dependence for the mean energy of the system $E_t = < p_t^2 > /2$ can be written in the form

$$E_t = \frac{1}{4\pi} \sum_k \Delta\varphi_k \left[J_k q \frac{t}{M_k} + f_k(t) \right]^2, \quad \sum_k \Delta\varphi_k = 2\pi, \qquad (10.68)$$

where J_k is the number of rotations, M_k is a period of the trajectory, and $f_k(t)$ is a periodic function with a period M_k. Note that in numerical experiments such blocks with different $\Delta\varphi_k$ are well observed.

From (10.68) it follows that if the number of trajectories with nonzero values of J_k is small, the behavior of the mean energy E_t at finite times will be similar to a quasi-periodic one. In the opposite case, the law of the increase of energy will have a strongly quadratic character.

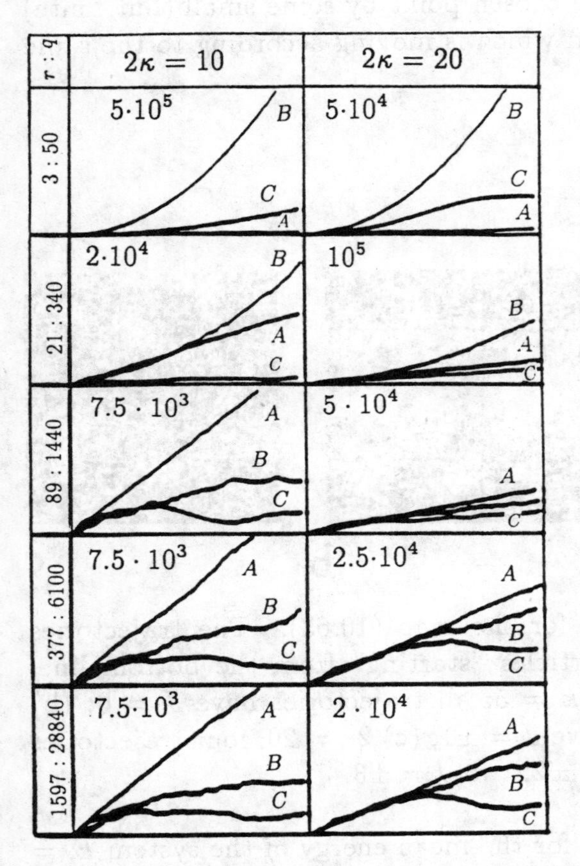

Fig. 16. Increase of the mean energy up to time $t = 200$ for classical rotator (curve A), classical model of quantum stochasticity (curve B) and quantum rotator (curve C). Numbers in figure show the scale of E_t along the vertical axis.

The number of particles with nonzero values of J_k will apparantly depend on the relation between the parameters of the system (10.67). Fig. 16 shows the law of increasing of the mean energy of the system described by the map (10.63) up to the time $t = 200$ for the values of parameters $2\kappa = 10,0$ and $2\kappa = 20$, and for different values of parameter ζ (curves B). The quantum parameter ζ was chosen to be rational $\zeta = r/q$ converging on the irrational number $(\sqrt{5}-1)/20$. Averaging was made over the segment $[0, 2\pi]$ which consists of a system of two thousand uniformly distributed particles. In order to control the evolution of the system we increased the averaging ensemble up to four thousand particles. This, practically, does not change the dependance E_t for the time-intervals under consideration. The numbers in the figure correspond to the maximal value of the mean energy on the vertical axis. It is seen from Fig. 16 that there occurs a change of a quadratic regime to that of "quantum diffusion limitation".

Now we turn to the problem of the time-scale τ_D of diffusion limitation. As we mentioned, the time τ_D has the meaning of the limitation time of classical diffusion for the quantum system (10.38) only when ζ is an irrational number. However, in numerical experiments one can use the resonant case $\zeta = r/q$ at sufficiently large r and q. As it is seen from Fig. 16, plots E_t for $\zeta = 377/6100$ and $\zeta = 1597/25340$ do not practically differ on the considered time-scale. This indicates the convergence of $E_t(\zeta_r)$ to $E_t(\zeta_{irr})$ at $\zeta_r = r/q \to \zeta_{irr}$ (existence of such a convergence for the kicked quantum rotator (10.38) is guaranteed by the estimate (10.41)).

It is evident that for our model with discrete phase space the time-scale τ_D is associated with the characteristic times of locking of trajectories. Information about a period of locking trajectories can be derived from the function $P(M)$ which itself represents the probability of unlocking for the trajectory within the time-interval M. In the numerical experiment the function $P(M)$ is calculated as the ratio of the number of trajectories which are not locked up to the time M to the total number of trajectories. The typical dependence $P(M)$ up to $M = 400$ is shown in Fig. 17. Numerical experiment shows that the form of $P(M)$ does not change when

$r, q \to \infty; r/q \to \zeta_{irr}$; i.e. at a given value of ζ, the function $P(M)$ depends mainly on κ.

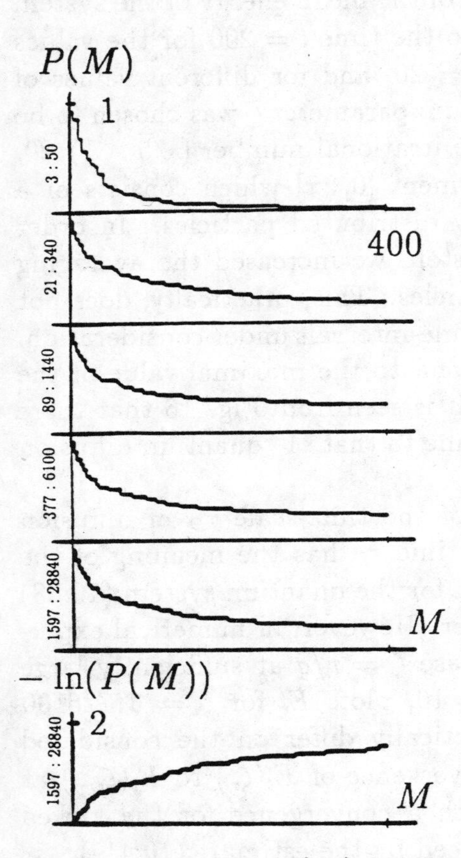

Fig. 17. Function $P(M)$ for the probability of unlocking trajectory for time M. Value of parameter $2\kappa = 10$.

When the dependence $P(M)$ is known it is easy to derive the upper estimate for the law of increasing of the mean energy E_t. Supposing that the energy increase of a particle for time $t < M$ (M is a period of locking) is determined by the diffusion law, and taking into account the fact that after times $t > M$ the particle under consideration does not contribute to the increase of E_t, we have

$$E_t < \kappa^2 \int_0^t P(M) dM. \tag{10.69}$$

Thus, the mean energy E_t deviates from the classical linear diffusion law, and by analogy with a quantum case it is possible to introduce a diffusion limitation time τ_D for the model (10.58), (10.63). As it is seen from Fig. 17, the probability function $P(M)$ has besides the characteristic time M_1^* (10.66) (which, actually, corresponds to the length of a "tail" of the function $P(M)$) another characteristic time M_2^* determining a sharp initial decrease of the function $P(M)$. The time M_2^* does not depend on q and determines, according to (10.69), the time-scale of diffusion limitation τ_D in the "classical model of quantum stochasticity".

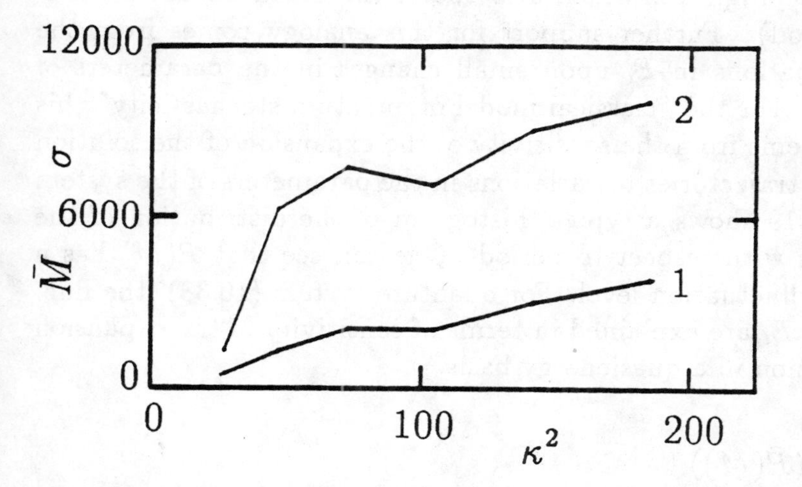

Fig. 18. Average closure period M (curve 1) and its standard deviation σ (curve 2) as functions on κ^2: ($\zeta = 3777/6100$; the average is taken over 6100 trajectories distributed uniformly in phase at $t = 0$ with $p_0 = 0$).

A rough estimate for time-scale M_2^* can be obtained if we assume that the number of accessible points in phase φ is of the order of the number of these points in p: $\sim \kappa\sqrt{2M_2^*}$. In this case we have for the effective size of phase space $N_2 \sim (\kappa\sqrt{2M_2^*})^2$ leading under the condition $M_2^* \sim \sqrt{N_2}$ to the estimate

$$M_2^* \sim \tau_D \sim 2\kappa^2. \tag{10.70}$$

The estimate (10.70) agrees with that of the diffusion limitation time-scale τ_D for the quantum kicked rotator [5]. We also note that the inequality (10.67) can be derived from the condition $M_1^* < M_2^*$.

We would also like to point out an analogy between an expansion of the solution of quantum system (10.38) in quasienergy states and an expansion of the solution of the "classical model of quantum stochasticity" (10.63) in periodic trajectories. The degree of delocalization of the quasienergy functions is associated with the expectation value $\bar{M}$ of the period of a trajectory. It is known that the degree of delocalization of quasienergy functions increases as κ^2 [5]. The analogous behaviour has been found for the average period $\bar{M}$ (see Fig. 18, which also shows the standard deviation σ of the period). Further support for this analogy comes from the large fluctuations in E_t upon small changes in the parameters of the system. For the "classical model of quantum stochasticity" this property stems from the sensitivity of the expansion of the solution in periodic trajectories to variations in the parameters of the system [91]. Fig. 19 shows a typical histogram of the distribution of the trajectories with respect to period. One can see that $\bar{P}(M)$ has a fairly high fluctuation level. For quantum system (10.38), the fluctuations in E_t are explained in terms of sensitivity of the expansion of the solution in a quasienergy basis.

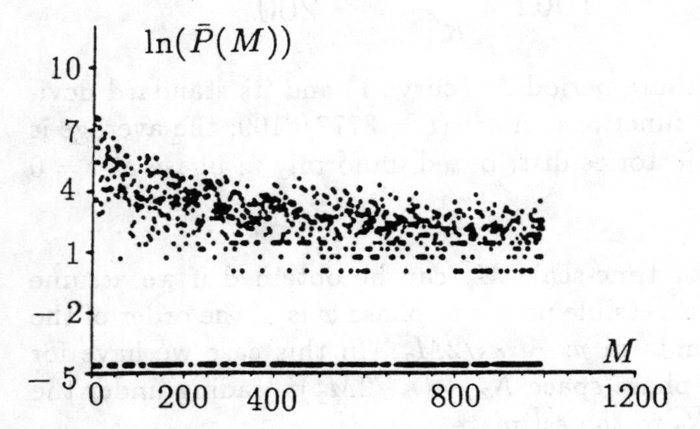

Fig. 19. Histogram distribution of the trajectory closure periods (logarithmic scale) for $\kappa = 10.1$ and $\zeta = 1597/25840$.

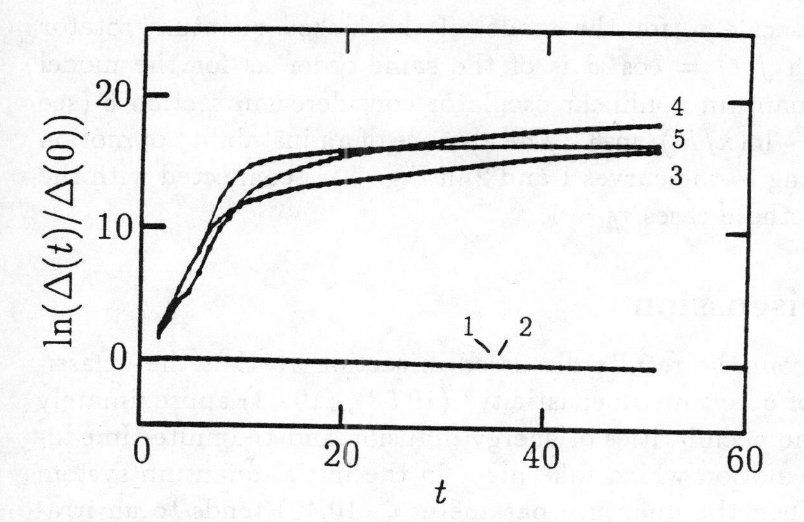

Fig. 20. Logarithm of the distance between two trajectories that are close together at $t = 0$, i.e., $\ln[\Delta(t)/\Delta(0)]$, averaged over 500 initial conditions; $[\Delta(0) = 1.1 \cdot 10^{-6}; K = 7.76; \zeta = (377/6100) \cdot 10^{-\alpha}; \kappa = 5 \cdot 10^{\alpha}]$; (1): $\alpha = 1$; (2): $\alpha = 2$; (3): $\alpha = 3$; (4): $\alpha = 4$; (5): $\alpha = \infty$ (classical limit).

Consider now the problem of how the logarithmically small time-scale $\tau_\hbar \sim \ln(1/\hbar)$ occurs in the "classical model of quantum stochasticity" which is described by the map (10.63). For this we investigate the stability of trajectories for model (10.63). Since the derivative of the function $[2\kappa \sin 2\varphi]$ with respect to φ is zero everywhere (except for a set of zero measure), the Lyapunov exponent for model (10.63) is equal to zero. Nevertheless, there is a certain instability of trajectories in the model (10.63). This instability is illustrated in Fig. 20, which is a logarithmic plot of the distance between two trajectories that were close at $t = 0$, for different values of the quasiclassical parameter κ. In the deep quasiclassical region the curves with ($\kappa \gg 1$) approach the classical curve with $\kappa = \infty$. The dynamics described by the map (10.63) for these values of κ corresponds during the finite times to the classical chaotic motion described by the map (10.47). The dynamics corresponding to the curves 1 and 2 should be considered as pure "quantum". Really, the estimate of the time-scale $\tau_\hbar$ for the model of the kicked quantum rotator (10.38) with $f(t) = \cos 2\theta$ is of the same order as for the model of kicked quantum nonlinear oscillator considered in section 6 (see (6.39)): $\tau_\hbar \sim \ln(\kappa/K)/\ln K$. The absence of an instability of motion corresponding to the curves 1 and 2 in Fig. 20 is connected with the fact that in these cases $\tau_\hbar \sim 1$.

10.7 Discussion

It follows from the results discussed in section 10 that the "classical model of quantum stochasticity" (10.58), (10.63) approximately describes the peculiarities of energy diffusion and the finite-time instability of motion which take place in the initial quantum system (10.38). When the quantum parameter ζ (10.40) tends to an irrational number ($r, q \to \infty; r/q \to \zeta_{irr}$), and the condition $2\kappa^2 \ll q$ is fulfilled, the model (10.63) reveals the characteristic limitation of diffusion of quantum mean energy in analogy with an initial quantum model (10.38); for both models the characteristic time-scales of diffusion limitation coincide in order of magnitude: $\tau_D \sim \kappa^2$.

It is important that the model (10.63) also describes a pure quantum phenomenon called "quantum resonance" [5,85], at which the mean energy has the characteristic asymptotic form t^2. In Fig. 16 such motion is observed at $q = 50$ and 340 for times $t \leq 200$. In the deep quasiclassical region ($\kappa \gg 1$) the model (10.63) includes the instability of trajectories in "quantum phase space" which is analogous to the property of local instability of trajectories in the corresponding classical model (10.38). For values of the parameter κ that are not large enough (when the time-scale $\tau_\hbar \sim 1$) the instability of "quantum trajectories" vanishes. The time-scale $\tau_\hbar$ is also investigated in the Q-representation in [9] for some models that are chaotic in the classical limit. It is shown in [9] that the characteristic time-scale where the classical local instability exists has the order: $\tau_\hbar \sim \ln(1/\hbar)/\lambda_L$, where λ_L is a characteristic Lyapunov exponent.

In conclusion we note that the approach discussed above for the study of dynamics of the kicked quantum rotator (10.38) is, in fact, a new variant of the quasiclassical perturbation theory. In this approach the trajectories are not the usual classical trajectories, but are the ones determined by the map (10.63). This allows one to take into account the property of discreteness of phase space, and to deal only with the periodic trajectories. The problem of an expansion of quantum expectation values in periodic trajectories is presently a topic of active research and is being discussed, in a variety of versions, in papers on quantum chaos (see, for example, [13,14] and references therein).

The "classical model of quantum stochasticity" can be constructed for various quantum models with kicked interaction. Generalization of this model for the case of many degrees of freedom does not present any difficulties. Further development of this model requires taking into account the quasirandom term in (10.48) (see also [92]), or the operator $\mathbf{G}_v$ in (10.62).

11 Quantum Chaos of Atoms in a Resonant Cavity

11.1 Introduction

The considerations given above show that for integrable systems, the time-scale $\tau_\hbar$ after which the quasiclassical approximation breaks down is of the order $\tau_\hbar \sim const/\hbar^\alpha$ (1.1) and is discussed already by Born [45] in connection with Einstein's critical remarks about quantum mechanics [43,44] (see also section 5.6). If chaotic dynamics appears in the classical limit of the system being considered, then in place of (1.1) we have $\tau_\hbar \sim const \cdot \ln(const/\hbar)$ (1.2), i.e., the time after which quantum corrections invalidate the classical description becomes much shorter than in the case (1.1).

This reduction of the time-scale $\tau_\hbar$ when the dynamics of the system is chaotic rather than integrable in the classical limit $\hbar = 0$ is a *precise formulation* of a phenomenon that one could attempt to detect experimentally. In this connection, the choice of a convenient model is very important. The quantum effects must be sensitive to variation of a parameter of the system that carries its classical dynamics from a regime of stable motion to a regime of stochastic motion.

The main aim of sections 11-14 is to describe two such systems and show that in these systems the effect of the transition of the time of breakdown of the quasiclassical approximation (1.1) to the time (1.2) occurs. These two systems are different nonintegrable generalizations of the integrable model which was introduced by Dicke [24] and bears his name.

In this section we consider, following the paper [20] (see also [21]), the system which consists of N atoms in a resonant cavity interacting self-consistently with an electromagnetic field within the cavity. Ideally, the atoms are treated as two-level systems. In what follows, we shall assume that the field has a single mode with frequency equal to the transition frequency of the atomic transition. In the resonance approximation, our model reduces to the Dicke model mentioned above. It is shown by Dicke [24] that the model permits the existence of a coherent super-radiant state, and a series of subsequent studies have made it possible to obtain various quantum solutions of the model, some of them exact [93]. This provides a basis for understanding many problems in the nonlinear dynamics of interacting fields and atoms. In 1976, a new version of the model was proposed [94]; in it, the nonresonant interaction term was taken into account explicitly. In the semiclassical approximation (classical field and two-level atoms), the entire problem can be effectively described as a classical model in which the dynamics of the two-level atoms is described by a variable of the type of classical angular momentum. In all that follows, when speaking of the classical or semiclassical limit, we mean the semiclassical approximation defined above. It was shown in [94] that, if the dimensionless coupling constant Λ of the atoms to the field reaches values of the order of unity, then the dynamics of the atomic variables becomes stochastic, and the field loses coherence. For $\Lambda \ll 1$, stochasticity exists only in an exponentially small region of phase space of the system. This coupling constant is given by

$$\Lambda = \sqrt{\frac{16\pi \rho_0 d_0^2}{\hbar\omega}}, \tag{11.1}$$

where $\rho_0 = N/V$ is the density of the atoms in the resonant cavity, d_0 is the dipole moment of the atomic transition being considered, and ω is the frequency of the atomic transition, equal to the frequency of the field. The condition of transition to strong chaos, $\Lambda > 1$, can be controlled by varying, for example, the density ρ_0. Subsequent investigations of the model proposed in [94] confirm the existence of chaos in it and a significant change of the quantum properties of the system on the passage through the value $\Lambda \sim 1$ [95-101]. We note that purely quantum investigations have been made only either numerically or in the approximation in which some quantum expectation values are decoupled.

A true quantum analysis requires us to give up the treatment of the field in the classical approximation and to take into account the corresponding quantum corrections. Their importance and, accordingly, the time-scale $\tau_\hbar$ of validity of the semiclassical approximation depends on whether or not there is strong chaotic dynamics in the model in the semiclassical limit; i.e., the quantum corrections depend on the value of the parameter Λ and on the extent to which the radiation field is semiclassical, measured by the value of $\kappa = I/\hbar$, where I is the classical value of the system's action action. It will be shown in this section that when Λ passes through the value $\Lambda \sim 1$ ($\Lambda > 1$) and $\kappa \gg 1$ the time-scale $\tau_\hbar$ becomes equal to $\tau_\hbar \sim \ln N$, where N is the number of atoms. Actually this result coincides with the estimate (1.2), as for the system considered here the quasiclassical parameter κ has an order: $\kappa = I/\hbar \sim N$. Herein, we shall make an investigation of the quantum model of [94] including the nonresonant terms, and numerically show the existence of exponentially rapid growth of the quantum corrections for the expectation values in the basis of coherent states. We also make an additional averaging over the initial density matrix (see sections 12 and 14). This exponential growth of the quantum corrections under the condition of chaos in the classical limit is what leads to the appearance of the logarithmic time-scale $\tau_\hbar \sim \ln N$. In deriving the equations for time-dependent expectation values in this problem we use the method which was discussed in section 3.

11.2 Quantum Equations of Motion

We represent the Hamiltonian of the system of N two-level atoms interacting with the radiation field in the resonant cavity in the form

$$\mathbf{H} = \hbar\omega[\mathbf{a}^{+}\mathbf{a} + \sum_{j=1}^{N} \mathbf{S}_j^z + \Lambda\frac{1}{\sqrt{N}}\sum_{j}^{N}\left(\mathbf{a}\mathbf{S}_j^{+} + \mathbf{a}^{+}\mathbf{S}_j^{-}\right) \tag{11.2}$$

$$+\Lambda\frac{1}{\sqrt{N}}\sum_{j}^{N}\left(\mathbf{a}\mathbf{S}_j^{-} + \mathbf{a}^{+}\mathbf{S}_j^{+}\right)],$$

where $\mathbf{a}^{+}$ and $\mathbf{a}$ are, respectively, the boson operators of creation and annihilation of the electromagnetic field; $\mathbf{S}_j^z, \mathbf{S}^{\pm} = \mathbf{S}_j^x \pm i\mathbf{S}_j^y$ are the operators of spin 1/2, describing the state of j-th two-level atom; ω is the frequency of the resonant mode in the cavity, equal to the frequency of atomic transition; and Λ is the dimensionless coupling constant defined in (11.1). The operators of the field and atoms satisfy the commutation relation

$$[\mathbf{a}, \mathbf{a}^{+}] = 1; \quad [\mathbf{S}_j^{+}, \mathbf{S}_j^{-}] = 2\mathbf{S}_j^z; \quad [\mathbf{S}_j^{\pm}, \mathbf{S}_j^z] = \mp\mathbf{S}^{\pm}. \tag{11.3}$$

The Hamiltonian (11.2) consists of four terms. The first and the second of them correspond to the field energy and to the energy of atoms, respectively. The third term

$$\mathbf{V}_r = \Lambda\frac{1}{\sqrt{N}}\sum_{j}^{N}\left(\mathbf{a}\mathbf{S}_j^{+} + \mathbf{a}^{+}\mathbf{S}_j^{-}\right) \tag{11.4}$$

corresponds to the resonant interaction of the atoms with the field, and the fourth term

$$\mathbf{V}_{ar} = \Lambda\frac{1}{\sqrt{N}}\sum_{j}^{N}\left(\mathbf{a}\mathbf{S}_j^{-} + \mathbf{a}^{+}\mathbf{S}_j^{+}\right), \tag{11.5}$$

describes the nonresonant interaction of the atoms with the field. The Dicke model corresponds to $\mathbf{V}_{ar} \equiv 0$, i.e., neglecting the nonresonant interaction. In the semiclassical approximation the complete problem (11.2) is considered in [94], where it is shown that chaos is due precisely to the presence of the term $\mathbf{V}_{ar}$.

In what follows, we shall be interested in the case when the mean field energy in the resonator,

$$E_f = \hbar\omega < \mathbf{a}^+\mathbf{a} >,$$

is comparable with the energy of the atoms,

$$E_a = \hbar\omega \sum_{j=1}^{N} < \mathbf{S}_J^z > .$$

More precisely, we shall assume that the following condition holds

$$E_f^{max} \sim E_a^{max} \propto N. \tag{11.6}$$

The condition (11.6) means that for the field the parameter governing its semiclassical behavior is

$$\epsilon = \frac{\hbar}{I_f^{max}} \sim \frac{1}{\kappa} \sim \frac{1}{N}, \tag{11.7}$$

i.e., for $N \gg 1$, the field is nearly in its classical state.

Using operators normalized to the number N of atoms,

$$\mathbf{A} = \frac{\mathbf{a}}{\sqrt{N}}, \quad \mathbf{A}^+ = \frac{\mathbf{a}^+}{\sqrt{N}}, \quad [\mathbf{A}, \mathbf{A}^+] = \frac{1}{N}, \tag{11.8}$$

$$\mathbf{S}^\nu = \frac{1}{N} \sum_j^N \mathbf{S}_j^\nu, \quad (\nu = z, \pm),$$

$$[\mathbf{S}^\pm, \mathbf{S}^z] = \mp \frac{1}{N} \mathbf{S}^\pm, \quad [\mathbf{S}^+, \mathbf{S}^-] = \frac{2}{N} \mathbf{S}^z,$$

recasts the Hamiltonian (11.2) in the form

$$\mathbf{H} = \hbar\omega N \left(\mathbf{A}^+\mathbf{A} + \mathbf{S}^z + \mathbf{V}_r + \mathbf{V}_{ar} \right), \tag{11.9}$$

$$\mathbf{V}_r = \Lambda \left(\mathbf{A}\mathbf{S}^+ + \mathbf{A}^+\mathbf{S}^- \right); \quad \mathbf{V}_{ar} = \Lambda \left(\mathbf{A}\mathbf{S}^- + \mathbf{A}^+\mathbf{S}^+ \right).$$

From (11.8) and (11.9) we derive the following operator equations of motion in the Heisenberg form

$$-i\frac{d\mathbf{A}^+}{d\tau} = \frac{1}{\hbar\omega}[\mathbf{H}, \mathbf{A}^+] = \mathbf{A}^+ + \Lambda(\mathbf{S}^+ + \mathbf{S}^-), \tag{11.10}$$

$$-i\frac{d\mathbf{S}^+}{d\tau} = \frac{1}{\hbar\omega}[\mathbf{H}, \mathbf{S}^+] = \mathbf{S}^+ - 2\Lambda(\mathbf{A}^+ + \mathbf{A})\mathbf{S}^z,$$

$$-i\frac{d\mathbf{S}^z}{d\tau} = \frac{1}{\hbar\omega}[\mathbf{H}, \mathbf{S}^z] = \Lambda(\mathbf{A}^+\mathbf{S}^- - \mathbf{A}\mathbf{S}^+) + \Lambda(\mathbf{A}\mathbf{S}^- - \mathbf{A}^+\mathbf{S}^+).$$

The equations for $\mathbf{A}$ and $\mathbf{S}^-$ are obtained from those for $\mathbf{A}^+$ and $\mathbf{S}^+$ by Hermitian conjugation, and $\tau = \omega t$. The system (11.10) contains both resonant and nonresonant interaction terms. If, in (11.9), we set $\mathbf{V}_{ar} \equiv 0$, then, in place of (11.10) we obtain the system of equations

$$-i\frac{d\mathbf{A}^+}{d\tau} = \mathbf{A}^+ + \Lambda\mathbf{S}^+, \tag{11.11}$$

$$-i\frac{d\mathbf{S}^+}{d\tau} = \mathbf{S}^+ - 2\Lambda\mathbf{A}^+\mathbf{S}^z,$$

$$-i\frac{d\mathbf{S}^z}{d\tau} = \Lambda(\mathbf{A}^+\mathbf{S}^- - \mathbf{A}\mathbf{S}^+),$$

which describes only the resonant interaction of the atoms with the radiation field, and corresponds to the Dicke model [24].

The normalization introduced in (11.8) makes it possible to give a very convenient interpretation of the quantum corrections to the problem described by the Hamiltonian (11.9). Since the dependence of $\mathbf{H}$ on N occurs only in the form of a factor, the equations of motion (11.10) and (11.11) do not depend explicitly on N. The same property is valid for the semiclassical limit of (11.10) and (11.11). Therefore, the dependence on N is manifested only for the quantum corrections, when the commutators in (11.8) are not identically equal to zero.

The system (11.11) is completely integrable. At the same time, in the semiclassical limit, $\mathbf{A}$ and $\mathbf{A}^+$ acquire the meaning of complex amplitudes of the field, $\mathbf{S}^{\pm}$ become complex amplitudes of the polarization of the atoms, and $\mathbf{S}^z$ becomes the difference of the populations of the atoms. In the case of (11.10), strongly stochastic dynamics arises in the semiclassical limit if $\Lambda > 1$, and the total energy $E_f + E_a \sim \hbar\omega N$. As seen below, allowance for the quantum corrections leads to a perturbation of the semiclassical approximation that has the order $1/N$ in the equations of motion for the quantum-mechanical expectation values. Therefore, in accodance with (1.2) we shall have for the time $\tau_\hbar$

$$\tau_\hbar \sim \ln N \tag{11.12}$$

by virtue of the estimate (11.7). This result will subsequently be confirmed numerically. Here we merely note that the number of atoms N appears as a reciprocal semiclassical parameter. Therefore, the problem of quantizing chaotic motion and the influence of quantum corrections is sensitive to a simple macroscopic parameter of the system.

11.3 Closed Equations of Motion for the Quantum-Mechanical Expectation Values

The operator equations (11.10) completely determine the evolution in time of all the operators. To solve these equations, it is also necessary to specify initial conditions. Equations (11.10) are very complicated, and a more convenient way of investigating them is to go over from the operator equations to the c-number equations that describe the dynamics of the quantum-mechanical expectation values of the variables (physical observables). To derive these equations, we note that, if equations (11.10) could be solved, then their solution would have the form of the following operator functions

$$\mathbf{A}^+(\tau) = \mathbf{A}^+(\mathbf{A}^+, \mathbf{A}, \mathbf{S}^z, \mathbf{S}^+, \mathbf{S}^-, \tau), \tag{11.13}$$

$$\mathbf{S}^\nu(\tau) = \mathbf{S}^\nu(\mathbf{A}^+, \mathbf{A}, \mathbf{S}^z, \mathbf{S}^+, \mathbf{S}^-, \tau), \quad (\nu = z, \pm),$$

where the operators $\mathbf{A}^+, \mathbf{A}, \mathbf{S}^z, \mathbf{S}^+, \mathbf{S}^-$ are given at $(\tau = 0)$.

We consider the expectation values $A(\tau)$, $A^+(\tau)$, $S^\nu(\tau)$, upon choosing a suitable basis for the perfoming the quantum-mechanical averaging. As a basis, we use wave functions of CSs of the field and atoms for $\tau = 0$ (see section 2). In addition, use of CSs makes it possible to obtain equations for $A(\tau)$, $A^+(\tau)$, $S^\nu(\tau)$ in closed form, in the same way as it was done in section 3.

It is convenient to define boson CSs $|z>$ as eigenvectors of the annihilation operator $\mathbf{A}$

$$\mathbf{A}|z>=z|z>. \tag{11.14}$$

Using the notation in (11.8) and the properties of boson CSs presented in section 2.2, we can write down the following relations for the boson CSs $|z>$ (11.14)

$$|z>=\exp(Nz\mathbf{A}^{+}-Nz^{*}\mathbf{A})|0>, \tag{11.15}$$

$$n(0)\equiv<z|\mathbf{a}^{+}\mathbf{a}|z>=N<z|\mathbf{A}^{+}\mathbf{A}|z>=N|z|^{2},$$

where $n(0)$ is the mean number of field photons at the initial time.

To introduce spin CSs at $\tau=0$, we first define the operator of the total spin

$$\mathbf{J}=\sum_{j=1}^{N}\mathbf{S}_{j}=N\mathbf{S}. \tag{11.16}$$

The operator $\mathbf{J}^{2}$ in an invariant

$$[\mathbf{J}^{2},\mathbf{H}]=0, \tag{11.17}$$

and the quantum number J corresponding to it (also called the co-operative number) is an integral of the motion. The values taken by J are: $J=0,...,N/2$. Using the usual representation for a wave function with given eigenvalues of the operator $\mathbf{J}^{2}$ and the operator $\mathbf{J}^{z}$, we have

$$\mathbf{J}^{2}|J,M>=J(J+1)|J,M>,\quad (M=-J,...,J). \tag{11.18}$$

The spin CS $|\xi>$ is introduced in the same way as discussed in section 2.3

$$|\xi>=\sum_{M=-J}^{J}U_{M}(\xi)|J,M>, \tag{11.19}$$

where, analogously to (2.55), we have for $U_{M}(\xi)$

$$U_{M}(\xi)=(1+|\xi|^{2})^{-J}\left[\frac{2J!}{(J+M)!(J-M)!}\right]^{1/2}\xi^{J+M}.$$

The representation of CS $|\xi>$ can be given analogously to (2.51)

$$|\xi>=(1+|\xi|^2)^{-J}\exp(\xi\mathbf{J}^+)|J,-J>, \qquad (11.20)$$

where $|\xi|$ varies in the interval $[0,\infty)$.

We define a CS of the complete system at $\tau=0$ as the product

$$|z,\xi>=|z>|\xi>, \qquad (11.21)$$

using the vectors $|z>$ and $|\xi>$ already introduced.

Let us consider now an arbitrary operator (see (3.53)) written in the Heisenberg representation

$$\mathbf{f}(t)\equiv\mathbf{f}\left(\mathbf{A}^+(t),\mathbf{A}(t),\mathbf{S}^z(t),\mathbf{S}^-(t),\mathbf{S}^-(t)\right). \qquad (11.22)$$

The initial condition $\mathbf{f}\equiv\mathbf{f}(t=0)$ for the operator $\mathbf{f}(t)$ is

$$\mathbf{f}\equiv\mathbf{f}(t=0)=\mathbf{f}\left(\mathbf{A}^+,\mathbf{A},\mathbf{S}^z,\mathbf{S}^-,\mathbf{S}^-\right). \qquad (11.23)$$

The time-evolution of the operator $\mathbf{f}(t)$ is determined by the Heisenberg equation

$$\frac{d\mathbf{f}}{d\tau}=\frac{i}{\hbar\omega}[\mathbf{H},\mathbf{f}]. \qquad (11.24)$$

To derive a closed c-number equation for the time-dependent expectation value

$$f(\tau)=<\xi,z|\mathbf{f}(\tau)|z,\xi>, \qquad (11.25)$$

we use the fact that in the case considered the Hamiltonian $\mathbf{H}$ (11.9) does not depend on time τ

$$\mathbf{H}(\tau)=\mathbf{H}(\tau=0)\equiv\mathbf{H}. \qquad (11.26)$$

We can now rewrite (11.24) in the form

$$\frac{d\mathbf{f}}{d\tau}=\frac{i}{\hbar\omega}[\mathbf{H},\mathbf{f}], \qquad (11.27)$$

and the averaging of (11.27) in CS $|z,\xi>$ gives

$$\frac{\partial f}{\partial\tau}=\frac{i}{\hbar\omega}<[\mathbf{H},\mathbf{f}]>. \qquad (11.28)$$

Naturally, equation (11.28) does not have a closed form. To obtain a closed form, we use the method described in section 3, taking into account that in the case considered here the formulae (2.47) and (2.65) should be replacsd by the following

$$< z|\mathbf{fA}^+|z >= \frac{1}{N} e^{-N|z|^2} \frac{\partial}{\partial z} e^{N|z|^2} f, \tag{11.29}$$

$$< z|\mathbf{A}^+\mathbf{f}|z >= \frac{1}{N} z^* f, \qquad < z|\mathbf{fA}|z >= \frac{1}{N} z f,$$

$$< z|\mathbf{Af}|z >= \frac{1}{N} e^{-N|z|^2} \frac{\partial}{\partial z^*} e^{N|z|^2} f,$$

$$< \xi|\mathbf{S}^+\mathbf{f}|\xi >= \frac{1}{N}(1+|\xi|^2)^{-2J} \left(2J\xi^* - \xi^{*2} \frac{\partial}{\partial \xi^*} \right) (1+|\xi|^2)^{2J} f,$$

$$< \xi|\mathbf{S}^-\mathbf{f}|\xi >= \frac{1}{N}(1+|\xi|^2)^{-2J} \frac{\partial}{\partial \xi^*} (1+|\xi|^2)^{2J} f,$$

$$< \xi|\mathbf{S}^z\mathbf{f}|\xi >= \frac{1}{N}(1+|\xi|^2)^{-2J} \left(\xi^* \frac{\partial}{\partial \xi^*} - J \right) (1+|\xi|^2)^{2J} f,$$

$$< \xi|\mathbf{fS}^+|\xi >= \frac{1}{N}(1+|\xi|^2)^{-2J} \frac{\partial}{\partial \xi} (1+|\xi|^2)^{2J} f,$$

$$< \xi|\mathbf{fS}^-|\xi >= \frac{1}{N}(1+|\xi|^2)^{-2J} \left(2J\xi - \xi^2 \frac{\partial}{\partial \xi} \right) (1+|\xi|^2)^{2J} f,$$

$$< \xi|\mathbf{fS}^z|\xi >= \frac{1}{N}(1+|\xi|^2)^{-2J} \left(\xi \frac{\partial}{\partial \xi} - J \right) (1+|\xi|^2)^{2J} f.$$

Expressions (11.29) enable us to reduce the right-hand side in (11.28), analogously to the results presented in section 3.6, to an expression that contains only the expectation value $f(\tau)$. Finally, we derive a closed c-number equation for $f(\tau)$ in the form (3.55), (3.56)

$$\frac{\partial f}{\partial \tau} = \left\{ \hat{K}_{cl} + \frac{1}{N} \hat{K}_q \right\} f, \tag{11.30}$$

where we have denoted

$$\hat{K}_{cl} = i\{ z^* \frac{\partial}{\partial z^*} - z \frac{\partial}{\partial z} + \Lambda \frac{J}{N} \frac{2}{(1+|\xi|^2)} \left(\xi^* \frac{\partial}{\partial z^*} - \xi \frac{\partial}{\partial z} \right) \tag{11.31}$$

$$+\Lambda\left(z^*\frac{\partial}{\partial\xi^*}-z\frac{\partial}{\partial\xi}\right)+\xi^*\frac{\partial}{\partial\xi^*}-\xi\frac{\partial}{\partial\xi}+\Lambda\left(z^*\xi^2\frac{\partial}{\partial\xi}-z\xi^{*2}\frac{\partial}{\partial\xi^*}\right)$$

$$+\Lambda\left[z\xi^2\frac{\partial}{\partial\xi}-z^*\xi^{*2}\frac{\partial}{\partial\xi^*}+z\frac{\partial}{\partial\xi^*}-z^*\frac{\partial}{\partial\xi}\right]$$

$$+\Lambda\left[\frac{J}{N}\frac{2}{(1+|\xi|)^2}\left(\xi\frac{\partial}{\partial z^*}-\xi^*\frac{\partial}{\partial z}\right)\right]\},$$

$$\hat{K}_q=-i\Lambda\left(\xi^{*2}\frac{\partial^2}{\partial z^*\partial\xi^*}-\xi^2\frac{\partial^2}{\partial z\partial\xi}\right)+i\Lambda\left(\frac{\partial^2}{\partial z^*\partial\xi^*}-\frac{\partial^2}{\partial z\partial\xi}\right). \quad (11.32)$$

Before discussing the form of the equations (11.30)-(11.32), we note that by setting the operator $\mathbf{f}(\tau)$ equal to $\mathbf{A}^+(\tau)$,$\mathbf{A}(\tau)$, $\mathbf{S}^z(\tau)$, $\mathbf{S}^+(\tau)$, $\mathbf{S}^-(\tau)$, in sucsession, we obtain a system of equations for the required expectation values $A^+(\tau)$, $A(\tau)$, A$S^z(\tau)$, $S^+(\tau)$, $S^-(\tau)$, each of which is to be regarded as a function of τ and subject to the initial conditions

$$A^+=z^*,\quad A=z, \quad (11.33)$$

$$S^z=-\frac{J}{N}\frac{1-|\xi|^2}{1+|\xi|^2},\quad S^+=\frac{J}{N}\frac{2\xi^*}{1+|\xi|^2},\quad S^-=\frac{J}{N}\frac{2\xi}{1+|\xi|^2}.$$

Let us recall some general properties of equations (11.30)-(11.32) that follow from the structure of the operators $\hat{K}_{cl}$ (11.31) and $\hat{K}_q$ (11.32) (see section 4). If in (11.30) we retain only the operator $\hat{K}_{cl}$, then equations (11.30) and (11.31) lead to the semiclassical model corresponding to the system (11.10) if all operators in (11.10) are taken to be c-numbers. All the quantum corrections in (11.30) are due to the second term, which, in accordance with (11.32), contains only second derivatives, which describe quantum dispersion. The formal transition to the semiclassical approximation in (11.30)-(11.32) is made by going to the limit

$$N\to\infty,\quad J\to\infty,\quad \frac{2J}{N}=g=const. \quad (11.34)$$

At the same time, we require here that the density of atoms $\rho_0=N/V=const.$ in (11.1) (see also section 13 where a discussion of this subject is given).

11.4 Quantum Correlation Functions

As in sections 3.7 and 8.4 we shall analyze the quantum corrections to the classical solutions by investigating the QCF (quantum correlation function) $P_{f,g}(\tau)$ of pairs of relevent operators. For two arbitrary operators $\mathbf{f}$ and $\mathbf{g}$, we have (see (8.20))

$$\frac{P_{f,g}(\tau)}{N} = < \mathbf{f}(\tau)\mathbf{g}(\tau) > - < \mathbf{f}(\tau) >< \mathbf{g}(\tau) > . \tag{11.35}$$

The quasiclassical parameter $1/N$ is introduced in (11.35) for convenience, as the right side in (11.35) is a pure quantum correction, and has the same order $1/N$. In the semiclassical limit (11.34), $P_{f,g}(\tau)/N$ vanishes at any time τ. We shall use below the QCF (11.35) to estimate the role of quantum corrections under conditions of regular and chaotic semiclassical dynamics.

The c-number differential equation for QCF $P_{f,g}(\tau)$ can be presented in the form

$$\frac{\partial P_{f,g}}{\partial \tau} = \hat{K}_{cl} P_{f,g} + \hat{K}_q fg - g\hat{K}_q f - f\hat{K}_q g + \frac{1}{N}\hat{K}_q P_{f,g}. \tag{11.36}$$

In deriving (11.36) we have used equations (3.60) and (11.30). Substitution (11.31) in (11.36) gives

$$\frac{\partial P_{f,g}}{\partial \tau} = \hat{K}_{cl} P_{f,g} + i\Lambda \left\{ (\xi^2 - 1)\frac{\partial^+}{\partial(z,\xi)} - (\xi^{*2} - 1)\frac{\partial^+}{\partial(z^*,\xi^*)} \right\} \times (f,g) + \frac{1}{N}\hat{K}_q P_{f,g}, \tag{11.37}$$

where we have used the notation

$$\frac{\partial^+(f,g)}{\partial(z,\xi)} \equiv \frac{\partial f}{\partial z}\frac{\partial g}{\partial \xi} + \frac{\partial f}{\partial \xi}\frac{\partial g}{\partial z}. \tag{11.38}$$

The initial condition for equation (11.37) is

$$P_{f,g}(\tau = 0) = 0. \tag{11.39}$$

Equation (11.37) has a form analogous to equation (8.22) derived for the spin system. All the terms in (11.37) are quantum corrections. The final term in (11.37) has an additional small factor $1/N$

and can be ignored in an approximate analysis of the solution of equation (11.37) during a finite time-interval. To determine the expectation values f and g, it is necessary to use equation (11.30) and the corresponding initial conditions. This closes the system of equations for determining the functions $P_{f,g}(\tau), g(\tau)$, and $f(\tau)$.

11.5 Breaking the Semiclassical Approximation

The difference between the exact values of the quantum-mechanical expectation values and their values in the semiclassical limit can be taken to be small or even zero at an initial time. However, with the passage of time this difference will grow. The aim of the present section is to derive the main equations which permit us to find the time $\tau_\hbar$ after which the difference becomes large. The method we use is analogous to that discussed in sections 4.1 and 8.5.

We use the following scheme of approximations to determine QCF $P_{f,g}(\tau)$. We calculate $f(\tau)$ ande $g(\tau)$ by means of (11.30) ignoring the terms of the order $1/N$, so

$$\frac{\partial f}{\partial \tau} = \hat{K}_{cl} f; \quad \frac{\partial g}{\partial \tau} = \hat{K}_{cl} g. \tag{11.40}$$

Substituting these values for f and g in (11.37), we have, up to terms of the next order in $1/N$

$$\frac{\partial P_{f,g}}{\partial \tau} = \hat{K}_{cl} P_{f,g} \tag{11.41}$$

$$+ i\Lambda \left\{ \xi^2 \frac{\partial^+}{\partial(z,\xi)} - \xi^{*2} \frac{\partial^+}{\partial(z^*,\xi^*)} - \mu \left(\frac{\partial^+}{\partial(z,\xi)} - \frac{\partial^+}{\partial(z^*,\xi^*)} \right) \right\} (f,g),$$

where for convenience we have introduced the symbol μ, such that $\mu = 0$ corresponds to the resonance approximation (i.e., $V_{ar} = 0$), and $\mu = 1$ corresponds to inclusion of V_{ar} and the possibility of strong chaos in the semiclassiacl limit (11.34) when $\Lambda > 1$.

To solve equations (11.41) for $P_{f,g}(\tau)$, we first find the functions $f(\tau)$ and $g(\tau)$ from (11.40) by the method of characteristics and then, by the same method, determine $P_{f,g}(\tau)$ from (11.41) with the calculated known functions $f(\tau)$ and $g(\tau)$ (see section 4.1, where a similar procedure is used). The result of the corresponding numerical analysis will be given in the next section.

Besides the QCF $P_{f,g}(\tau)$ we can also find the dynamics of the quantum-mechanical expectation values. For this, we turn to the system of operator equations (11.10). We use the new system of operators

$$\mathbf{A}^{+} = \mathbf{A}^{x} + i\mathbf{A}^{y}, \quad \mathbf{S}^{\pm} = \mathbf{S}^{x} \pm i\mathbf{S}^{y}, \tag{11.42}$$

where $\mathbf{A}^{x,y}$ and $\mathbf{S}^{x,y}$ are Hermitian operators. Then, projection of the system (11.10) onto the CS $|z, \xi>$ gives

$$\frac{dA^{x,y}}{d\tau} = \mp A^{y,x} + \Lambda(\mu \mp 1)S^{y,x}, \tag{11.43}$$

$$\frac{dS^{x,y}}{d\tau} = \mp S^{y,x} \pm 2(1 \mp \mu)\Lambda A^{y,x} \cdot S^{z} \pm \frac{2}{N}(1 \mp \mu)\Lambda P_{A^{y,x},S^{z}},$$

$$\frac{dS^{z}}{d\tau} = 2\Lambda\{(\mu - 1)A^{y} \cdot S^{x} + (\mu + 1)A^{x} \cdot S^{y} + \frac{1}{N}(\mu - 1)P_{A^{y},S^{x}}$$

$$+ \frac{1}{N}(\mu + 1)P_{A^{x},S^{y}}\}.$$

The system of equations (11.43) is exact and actually completes the closure procedure if, for QCFs $P_{f,g}(\tau)$, we use equation (11.37). In what follows, we shall use the approximate equation (11.41) for the QCF $P_{f,g}(\tau)$.

In the general case ($\mu = 1$) of the system with the Hamiltonian (11.9), there are only two integrals of the motion: $\mathbf{H}$ and

$$\mathbf{J}^2 = N^2\{(\mathbf{S}^{x})^2 + (\mathbf{S}^{y})^2 + (\mathbf{S}^{z})^2\}, \tag{11.44}$$

(see (11.17)). In the resonance approximation ($\mu = 0$), corresponding to the Dicke model, there is an additional third integral of motion

$$\mathbf{W} = \mathbf{A}^{+}\mathbf{A} + \mathbf{S}^{z}, \tag{11.45}$$

which makes it possible to integrate the problem.

In the process of numerical analysis, the accuracy of the calculations is tested by verifying the conservation of the expectation values of the integrals of motion, in particular, the dimensionless energy density

$$\rho_E(\tau) \equiv \frac{H}{N\omega} = (A^x)^2 + (A^y)^2 + \frac{1}{N}(P_{A^x,A^x} + P_{A^y,A^y}) + S^z \quad (11.46)$$

$$+2\Lambda\{(1+\mu)A^x \cdot S^x + (1-\mu)S^y + \frac{1}{N}(1+\mu)P_{A^x,S^x}$$

$$+\frac{1}{N}(1-\mu)P_{A^y,S^y}\} = \rho_E(0),$$

where we have used the definition for an arbitrary expectation value: $f(\tau) \equiv < \xi, z|\mathbf{f}(\tau)|z, \xi >$. The values of quantum corrections were estimated in the same way as in section 8.5. Consider the "vector of variables" $\vec{D}(\tau) = (A^x, A^y, S^x, S^y, S^z)$. Its Euclidian length is

$$D(\tau) = \sqrt{(A^x)^2 + (A^y)^2 + (S^x)^2 + (S^y)^2 + (S^z)^2}. \quad (11.47)$$

In the limit (11.34), the expression (11.47) determines the semiclassical vector $\vec{D}_{cl}(\tau)$. The quantity

$$\delta(\tau) = |\vec{D}(\tau) - \vec{D}_{cl}(\tau)|/D_{cl}(\tau) \quad (11.48)$$

is what we construct numerically, and this determines the relative magnitude of the quantum corrections. In what follows, we assume the initial condition $\delta(\tau = 0) = 0$ so the initial condition is a semiclassical state. The time-scale $\tau_\hbar$ of the breaking of the semiclassical approximation is determined from the condition (see (8.49))

$$\delta(\tau_\hbar) = const. \quad (11.49)$$

The choice of the constant in (11.49) is nominal. Below, we assumed $const. = 0.02$, whereas N reached values $N_{max} \sim 10^5$.

11.6 Results of Numerical Analysis

As initial conditions the following were chosen

$$A^x(0) = A^y(0) = 1, \quad S^x(0) = 0.25, \quad S^y(0) = 0.05, \quad S^z(0) = 0.43. \quad (11.50)$$

Fig. 21 shows the dependence $S^z(\tau)$ in the resonant approximation ($\mu = 0$), and in the semiclassical limit. In this case, there are nonlinear periodic oscillations with cooperative frequency [24]. In the case of small oscillations, this cooperative frequency is equal to $\omega_c = \Lambda\omega$. Fig. 22 gives the dependence of the time $\tau_\hbar$ on the number N in the same resonance case ($\mu = 0$), when the problem can be exactly integrated. We calculate $\tau_\hbar$ in accordance with (11.49) with *const.* $= 0.02$. The dependence of $\tau_\hbar$ on N is a power law in the region of large N. Such behaviour of $\tau_\hbar$ as a function of the quasiclassical parameter (see (11.7)), agrees with the previously discussed results for integrable systems (see section 5). The dynamics of the correlation function

$$P(\tau) = |P(A^x, S^y)| + |P(A^y, S^x)| + |P(A^x, S^y)| \qquad (11.51)$$
$$+|P(A^x, S^z)| + |P(A^y, S^z)|,$$

corresponding to the given case ($\mu = 0$), is shown in Fig. 23. It can be seen that $P(\tau)$ increases slowly with time (as a power).

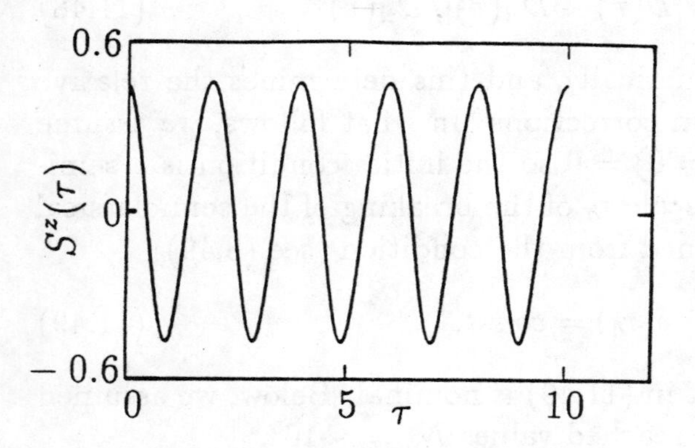

Fig. 21. Time depemdence $S^z(\tau)$ in the semiclassical limit ($\tau = \omega t$): integrable case ($\mu = 0$); $\Lambda = 1$. In all the subsequent figures in section 11, the parameters and initial conditions (unless specially stated otherwise) are the same.

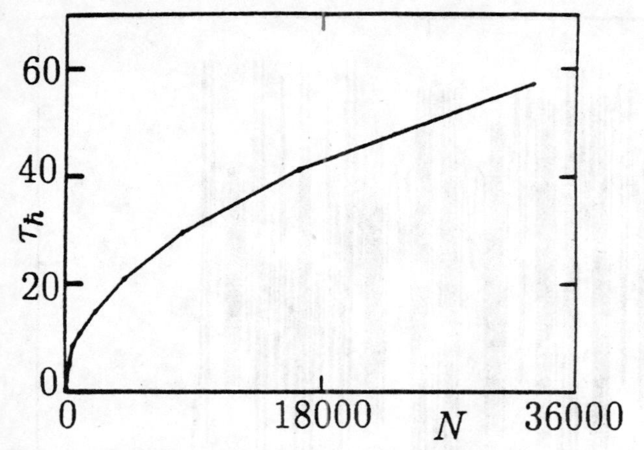

Fig. 22. Dependence of the time $\tau_\hbar$ on the number of atoms N (quasiclassical parameter of the field) in the integrable case ($\mu = 0$). The dependence $\tau_\hbar(N)$ was calculated in accordance with (11.49) with *const.* = *0.02*.

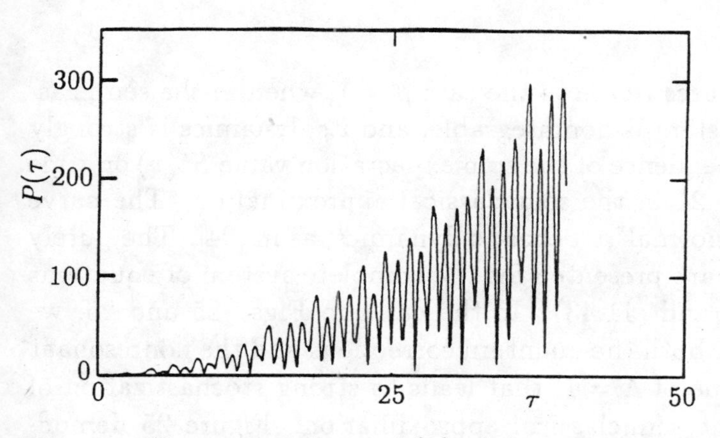

Fig. 23. Time dependence $P(\tau)$ of the quantum correlation function (11.51) in the integrable case ($\mu = 0$). The true quantum correlations are $P(\tau)/N$.

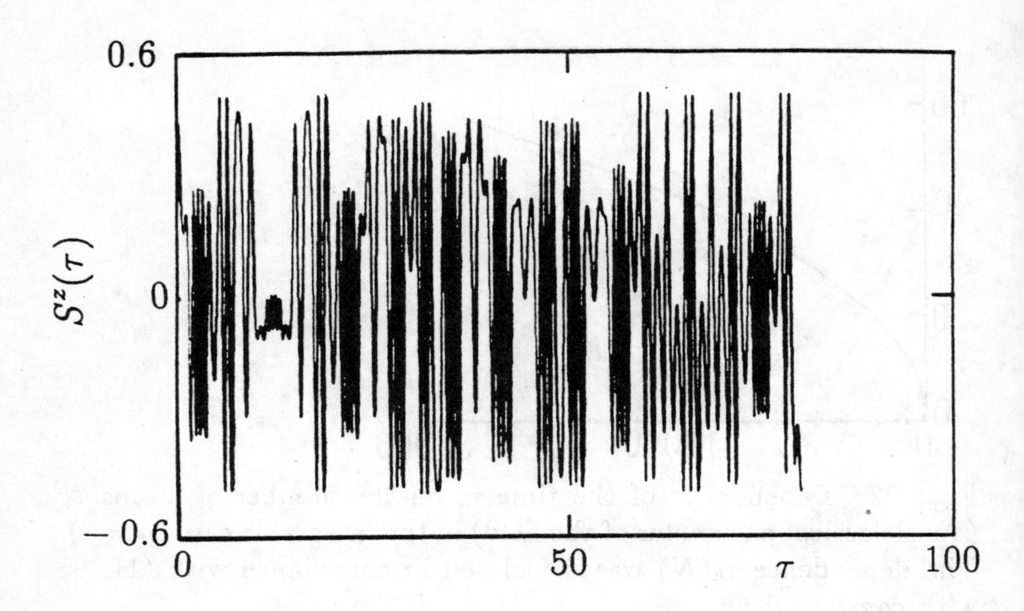

Fig. 24. The dependence $S(\tau)$ for the case of semiclassical chaotic motion ($\mu = 1$).

The following figures relate to the case $\mu = 1$, when in the semiclassical limit the system is nonintegrable, and its dynamics is strongly chaotic. The dependence of the spin expectation value $S^z(\tau)$ on time is shown in Fig. 24 in the semiclassical approximation. The curve corresponds to "normal" stochastic dynamics, as in [94]. The purely quantum results are presented for the complete system of equations (11.40), (11.41), and (11.43). In the cases in Figs. 25 and 26, we take into account both the quantum corrections and the nonresonant terms with a value of $\Lambda = 1$, that leads to strong stochastization of the motion in the semiclassical approximation. Figure 25 demonstrates the logarithmic dependence of $\tau_\hbar$ on N in accordance with the expressions (1.2) and (11.7)

$$\tau_\hbar \sim const. \ln(const. N), \tag{11.52}$$

where the values of the constants can depend on the kind of the problem being considered (see, in particular, [1] and section 6.3). We note, that N reaches very large values$\sim 10^5$.

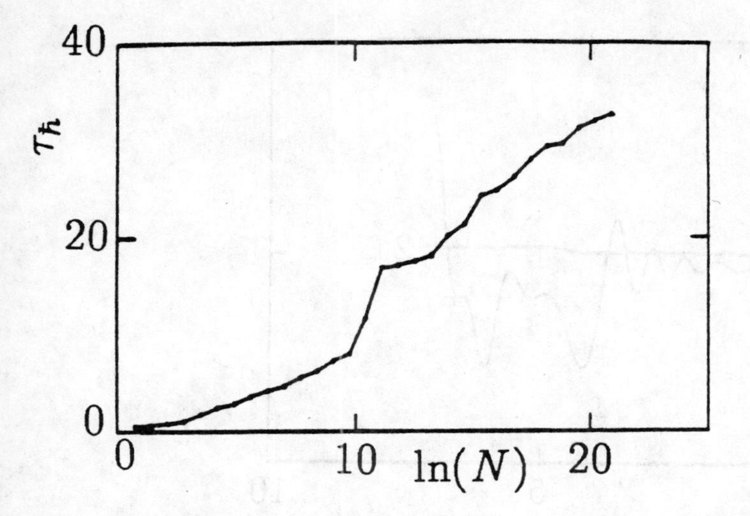

Fig. 25. Dependence $\tau_\hbar$ on N for the case of semiclassical chaotic motion.

Simultaneously with result (11.52), the correlation function $P(\tau)$ from (11.51) increases exponentially with the time (Fig. 26).

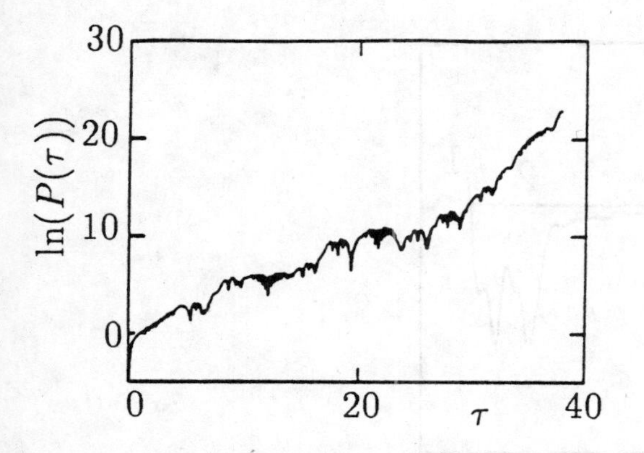

Fig. 26. Time dependence $P(\tau)$ of the quantum correlation function (11.51) in the case of chaotic semiclassical motion.

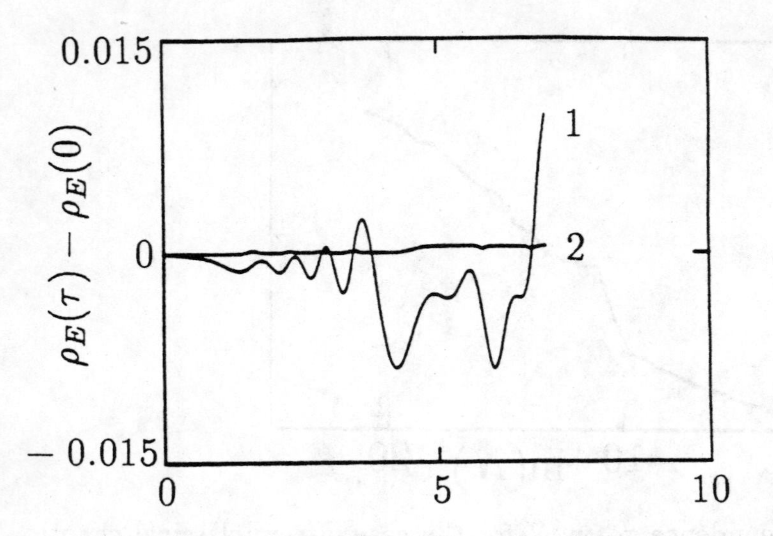

Fig. 27. Verification of the conservation of an integral of the motion - the dimensionless energy $\rho_E(\tau)$ (11.46) - for $N = 8192$. Curve 1 corresponds to a calculation in accordance with (11.46) without allowance for quantum correlation functions. Curve 2 corresponds to allowance for them in (11.46).

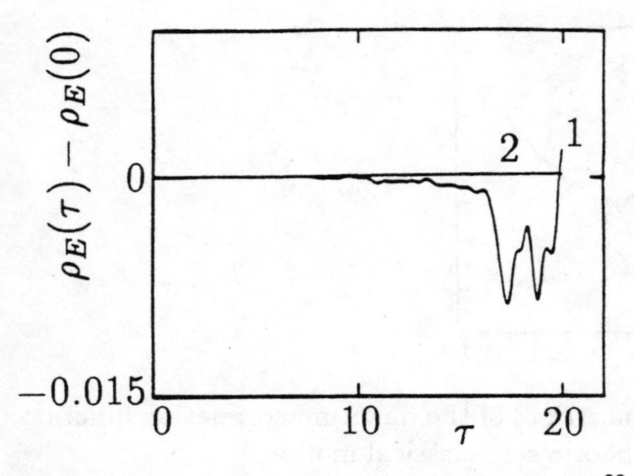

Fig. 28. The same as in Fig. 27 for $N = 2^{20}$.

As we have already noted, the logarithmic dependence of $\tau_\hbar$ on N is an important physical result, which, in general, can also be extended to other cooperative models, which will be discussed in sections 12-14. In fact, the assertion of the existence of the dependence (11.52) is the main result of this section. The dependence (11.52) can be experimentally verified, for example, by measuring the behaviour in time of the corresponding quantum correlation functions. As comparison of Figs. 26 and 23 shows, there is an important difference between the behavior in time of the quantum correlations in the integrable ($\mu = 0$) and nonintegrable ($\mu = 1$) cases. To conclude this section, we give the results of a verification of the accuracy of application of the scheme of perturbation theory explained at the end of section 11.4. One test of the accuracy is realized by calculating the expectation value of the energy density of the system in accordance with the expression (11.46). Figures 27 and 28 give the results of numerical calculation of the difference of the densities $\rho_E(\tau)$ at the time τ and $\rho_E(0)$ at the initial time. In an exact calculation, this difference would be equal to zero, since the system (11.9) is conservative. In Fig. 27, curve 1 corresponds to the calculation of $\rho_E(\tau)$ without allowance for the quantum correlation functions in (11.46). It can be seen that $\rho_E(\tau)$ is not conserved even at short times. Curve 2 in Fig. 27 corresponds to allowance for the contribution of the correlation functions to (11.46) in the first order in the small parameter $1/N$. The conservation of the energy density $\rho_E(\tau)$ in this case is satisfactory.

Figure 28 differs from the previous figure by an increase in the number of atoms to $N = 2^{20} \approx 10^6$. It can be seen that allowance for the correlation functions (curve 2) leads to a more accurate conservation of the energy density $\rho_E(\tau)$ because of the smaller value of quantum corrections.

11.7 Conclusion

In this section, we have given a theory for obtaining the quantum corrections for the description of the dynamics of a system of atoms interacting with a radiation field in a resonant cavity. The basis of the method is the derivation in closed form of equations of motion for the quantum-mechanical expectation values and quantum correlation functions. In the resonant approximation, the model corresponds to the well-known Dicke model [24], while in the complete form (i.e., with the nonresonant terms) it admits classical dynamics with strong chaos [94]. The numerical analysis indicates a strong difference in the nature of the growth of the quantum corrections depending on whether the problem is integrable or stochastic in the classical limit. The characteristic time-scale of breakdown of the semiclassical approximation is simply the logarithm of the (large) number N of atoms, and this circumstance is particularly important in the investigation of finite macroscopic systems.

12 Quantum Chaos of Atoms in a Resonant Cavity Driven by an External Resonant Field

12.1 Introduction

In the previous section we discussed the quantum "ln N-effect" under the condition of strong semiclassical chaos ($\Lambda > 1$) in the system with the Hamiltonian (11.2). Actually, in an experimental realization in the optical and even in the microwave frequency regions the condition $\Lambda > 1$ is difficult to achieve. That is why, below, in this section we consider, following [22], a modification of the quantum system with the Hamiltonian (11.2), where strong semiclassical chaos and the "ln N-effect" discussed in section 11 can be realized under the condition of weak interaction of atoms with the radiation field ($\Lambda \ll 1$).

As before, our system consists of an ensemble of N identical two-level atoms in a resonator, interacting with a single resonant eigenmode, which is assumed to be homogeneous. Besides, we assume that atoms are interacting with an additional external coherent field with a frequency slightly different from a frequency of atomic transition, and with a constant amplitude. Since this problem involves only finite times in the range $0 < \tau < \tau_\hbar$, in which the difference between classical and quantum dynamics is to be observed ($\tau \sim \tau_\hbar$), we shall not take into account effects of dissipation and effects connected with finite temperature. Thus, as in section 11, the considerations below will be given in the framework of the Hamiltonian approach. The system considered in this section possesses some useful advantages. Namely, as shown in [102] in the semiclassical approximation the transition to the global dynamical chaos takes place in this system at rather easily achieved conditions, which roughly can be presented as

$$\omega_c \sim \omega_R \sim |\Delta|, \tag{12.1}$$

where ω_c is the cooperative frequency ; ω_R is the Rabi frequency of slow oscillations of an individual atom in the mentioned above external resonant field; Δ is the difference between the frequencies of the external field and the atomic transition. The problems connected with occurence of dynamical chaos in the vicinity of a separatrix in the semiclassical approximation are considered in [103,104]. The quasiclassical parameter in this system is the same as in the system (11.2), $\epsilon = 1/N$, and since the number of atoms N may be rather large, there exists a possibility to study quantum dynamics under the conditions of developed chaos in the quasiclassical limit in the "deep" quasiclassical region ($\epsilon \ll 1$). In the computer calculations presented below N was chosen as $N < N_{max} \sim 10^9$ ($\epsilon_{min} \sim 10^{-9}$).

In this section we show that developed chaos in the semiclassical limit takes place under conditions $\Lambda \ll 1$ that could be interesting from the point of view of the experimental verification of the results discussed. Then, as in the previous section, the time-scale $\tau_\hbar$ is analyzed, and it is shown numerically that $\tau_\hbar$ is of the order: $\tau_\hbar \sim \ln N$. Also, quantum correlation functions are shown to grow exponentially on the time interval: $0 < \tau < \tau_\hbar$.

This section is organized in the following way. In section 12.2 we introduce the model, and construct its Hamiltonian in collective operators which describe the slow quantum dynamics. In section 12.3 two different approaches are given for deriving time-independent Hamiltonians for our model, which simplify significantly the further analysis. In section 12.4 an exact c-number equation is derived for time-dependent quantum expectation values in spin and boson coherent states. It is shown that the two approaches considered in section 12.3 lead to the same c-number equations. In section 12.5 some characteristic properties of the semiclassical approximation are discussed under the conditions of regular and chaotic dynamics. The role of quantum effects is considered in section 12.6 under the condition of developed chaos in the semiclassical limit. In section 12.7 an additional averaging over the initial density matrix is considered. Section 12.8 discusses the results.

12.2 The Formulation of the Model

Consider, as in the previous section, a system of N two-level atoms in a resonator. These atoms interact selfconsistently by dipole interaction with a single quantised eigenmode, which will be described as a quantum linear oscillator. Besides, we shall assume that an external coherent electro-magnetic field with a frequency slightly different from the frequency of atomic transition is injected into the resonator. The amplitude of the external field is assumed to be constant. In this case the Hamiltonian of the system can be written in the form

$$\mathbf{H} = \mathbf{H}_0 + \mathbf{V}_{int}(t), \tag{12.2}$$

$$\mathbf{H}_0 = \hbar\omega \mathbf{a}^+\mathbf{a} + \hbar\omega \sum_{j=1}^{N} \mathbf{S}_j^z,$$

$$\mathbf{V}_{int}(t) = \sqrt{\frac{2\pi d_0^2 \hbar\omega}{V}}(\mathbf{a}^+ + \mathbf{a}) \sum_{j=1}^{N} (\mathbf{S}_j^+ + \mathbf{S}_j^-) + 2\gamma \cos(\Omega t) \sum_{j=1}^{N} (\mathbf{S}_j^+ + \mathbf{S}_j^-).$$

In (12.2) the operators $\mathbf{a}^+$ and $\mathbf{a}$ describe the eigenmode of a resonator with frequency ω; the operator

$$\mathbf{S}_j \equiv \left(\mathbf{S}_j^z, \mathbf{S}_j^+, \mathbf{S}_j^-\right), \quad (j = 1, ..., N)$$

describes $j-th$ two-level atom with the frequency ω (for simplicity we have chosen here and in what follows the frequency of the eigenmode to be equal to the frequency of the atomic transition); d_0 and V are the dipole moment of the atomic transition and the volume of the resonator, respectively (the volume V can be considered also as an effective volume, where the interaction of atoms with the radiation field takes place [25]); γ and Ω are the amplitude and the frequency of the external coherent field. The interaction constant chosen in (12.2) is connected with quantization in a one-mode ring resonator (see [102-104]). Commutation relations for operators $\mathbf{a}^+$, $\mathbf{a}$, and $\mathbf{S}_j$ are given by formulae (11.3).

As already mentioned in the Introduction, our approach with the Hamiltonian (12.2) does not take into consideration many effects, especially of the relaxation type. The role of these effects can be rather small in the experiments connected with the model (12.2), especially in a regime of global chaos, where quantum effects exhibit themselves at small enough times $\tau < \tau_\hbar \sim \ln N$, and can be described in the framework of Hamiltonian approach.

We would like to apply to the system with Hamiltonian (12.2) the method based on the exact closed c-number equations for quantum dynamical expectation values in spin and boson coherent states given at $\tau = 0$, and discussed in detail in section 11. There we mentioned that this approach possesses two useful properties: 1) The c-number equations are derived directly for expectation values, but not, as usual, for the density matrix. Moreover, these equations separate in a very natural way into semiclassical (classical) and quantum parts. This property is very convenient when considering the influence of quantum corrections on special semiclassical solutions (partialarly when the semiclassical solution is chaotic). 2) Although these equations describe the dynamics only in coherent states, this circumstance does not present any limitations, since the dynamics of arbitrary expectation values can be completely defined from the dynamics of diagonal matrix elements in coherent states, by an additional averaging over the initial density matrix [30] (see also section 12.7).

However, the method used in section 11 can be applied directly only to Hamiltonians that have no explicit dependence on time. Since the Hamiltonian (12.2) includes an explicit dependence on time, this method can not be directly used in this case. We consider here two different approaches to the system with Hamiltonian (12.2) that allow one to derive c-number equations for quantum expectation values. Also we show that both these approaches lead to the same c-number equations. The approaches considered below also could be used in more general cases, in which the operator equations include an explicit dependence on time.

First, however, we shall simplify the Hamiltonian (12.2). To do this, we shall transform to the interaction representation with the Hamiltonian $\mathbf{H}_0$ in (12.2). In this representation the dependence on time of an arbitrary operator $\mathbf{f}$ has the form

$$\mathbf{f}(t) = e^{i\mathbf{H}_0 t/\hbar}\mathbf{f}e^{-i\mathbf{H}_0 t/\hbar}. \tag{12.3}$$

In particular, using $\mathbf{H}_0$ in (12.2) gives

$$\mathbf{a}(t) = e^{-i\omega t}\mathbf{a}, \quad \mathbf{a}^+(t) = e^{i\omega t}\mathbf{a}^+, \tag{12.4}$$

$$\mathbf{S}_j^z(t) = \mathbf{S}_j^z, \quad \mathbf{S}_j^-(t) = e^{-i\omega t}\mathbf{S}_j^-, \quad \mathbf{S}_j^+(t) = e^{i\omega t}\mathbf{S}_j^+.$$

Upon using (12.4), the Schödinger equation for the wave function $\Psi_i(t)$

$$\Psi(t) = \exp\left(-\frac{i}{\hbar}\mathbf{H}_0 t\right)\Psi_i(t), \tag{12.5}$$

in the interaction representation becomes

$$i\hbar\frac{\partial\Psi_i(t)}{\partial t} = \mathbf{H}_{int}(t)\Psi_i(t), \tag{12.6}$$

where the Hamiltonian $\mathbf{H}_{int}(t)$ has the form

$$\mathbf{H}_{int}(t) = e^{i\mathbf{H}_0 t/\hbar}\mathbf{V}_{int}e^{-i\mathbf{H}_0 t/\hbar} \tag{12.7}$$

$$= \sqrt{\frac{2\pi d_0^2\hbar\omega}{V}}\sum_{j=1}^{N}\left[(\mathbf{a}\mathbf{S}_j^+ + \mathbf{a}^+\mathbf{S}_j^-) + (\mathbf{a}\mathbf{S}_j^- e^{-2i\omega t} + \mathbf{a}^+\mathbf{S}_j^+ e^{2i\omega t})\right]$$

$$+2\gamma\cos(\Omega t)\sum_{j=1}^{N}(e^{i\omega t}\mathbf{S}_j^+ + e^{-i\omega t}\mathbf{S}_j^-).$$

In what follows we assume that the frequency Ω of the external field is close (but not equal) to the frequency of the atomic transition ω

$$\Omega - \omega = \Delta, \quad |\Delta| \ll \omega, \Omega. \tag{12.8}$$

Actually, condition (12.8) means the external field interacts resonantly with the atomic sub-system. In addition, we shall assume that the dimensionless constant of interaction of the atoms with the radiation field Λ_0 is small

$$\Lambda_0 = \sqrt{\frac{2\pi \rho_0 d_0^2}{\hbar\omega}} \ll 1, \tag{12.9}$$

where $\rho_0 = N/V$ is the density of atoms in the resonator (or the number of atoms in an effective volume of interaction). To avoid any misunderstanding, we note that Λ_0 introduced in (12.9) differs by the factor $\sqrt{8}$ from the constant Λ (11.1), used in section 11. The reason is that traveling waves are chosen for the quantization of the radiation field in this section.

Under the condition (12.9) the relative contribution of the fast motion in the full dynamics is small ($\sim \Lambda_0$), so we may consider only the slow dynamics. From (12.7) we derive the Hamiltonian that describes only the slow dynamics of atoms and field

$$\mathbf{H}_{int}^{sl}(t) \equiv \hbar\omega\Lambda_0\tilde{\mathbf{H}}_{int}(t) \tag{12.10}$$

$$= \hbar\omega\Lambda_0 N\left[(\mathbf{A}\mathbf{S}^+ + \mathbf{A}^+\mathbf{S}^-) + \lambda(e^{i\Delta t}\mathbf{S}^- + e^{-i\Delta t}\mathbf{S}^+)\right],$$

where the dimensionless constant λ

$$\lambda = \frac{\gamma}{\hbar\omega\Lambda_0} \tag{12.11}$$

describes the interaction of the atomic sub-system with the external field. The Hamiltonian (12.10) is expressed in terms of the collective operators introduced in (11.8).

The wave function $\tilde{\Psi}_i(\tau)$ describes the slow dynamics of our system in the interaction representation, and satisfies to the Schödinger equation

$$i\frac{\partial\tilde{\Psi}_i(\tau)}{\partial\tau} = \tilde{\mathbf{H}}_{int}(\tau)\tilde{\Psi}_i(\tau), \tag{12.12}$$

with the following Hamiltonian

$$\tilde{\mathbf{H}}_{int}(\tau) = N\left[(\mathbf{A}\mathbf{S}^+ + \mathbf{A}^+\mathbf{S}^-) + \lambda(e^{i\bar{\Delta}\tau}\mathbf{S}^- + e^{-i\bar{\Delta}\tau}\mathbf{S}^+)\right], \tag{12.13}$$

where we define

$$\bar{\Delta} = \Delta/\omega_c, \quad \tau = \omega_c t, \tag{12.14}$$

$$\omega_c = \omega\Lambda_0 = \sqrt{\frac{2\pi\rho_0 d_0^2\omega}{\hbar}},$$

$$(\omega_c/\omega = \Lambda_0 \ll 1).$$

The frequency ω_c is a cooperative frequency [24] which characterizes the time-scale of slow nondissipative oscillations connected with the interaction of the atomic and radiation field sub-systems.

Note that the Hamiltonian (12.13) transforms at $\lambda = 0$ to the quantum version of the Dicke model (see, for example, [24,105], where $\omega_c = k\sqrt{N}$).

The Hamiltonian $\tilde{\mathbf{H}}_{int}(\tau)$ also depends explicitly on time, and the method used in section 11 can not be applied directly. In the next section we consider two variants which allow one to derive time-independent Hamiltonians from (12.13).

12.3 Time-Independent Hamiltonian for Slow Dynamics

A. Direct method connected with a transformation to a rotating system of coordinates.

The simplest way to make the Hamiltonian (12.13) time -independent consists in transforming to a system of coordinates which rotates with the frequency $\bar{\Delta}$. For this introduce the new operators

$$\boldsymbol{\eta}^+ = e^{-i\bar{\Delta}\tau}\mathbf{S}^+, \quad \boldsymbol{\eta}^z = \mathbf{S}^z, \quad \mathbf{c} = e^{i\bar{\Delta}\tau}\mathbf{A}. \tag{12.15}$$

This transformation does not change the commutation relations (11.8), and leads to the following equations of motion with the constant coefficients

$$\dot{\mathbf{c}} = i\bar{\Delta}\mathbf{c} - i\boldsymbol{\eta}^-, \tag{12.16}$$

$$\dot{\boldsymbol{\eta}}^- = i\bar{\Delta}\boldsymbol{\eta}^- + 2i\boldsymbol{\eta}^z\mathbf{c} + 2i\lambda\boldsymbol{\eta}^z,$$

$$\dot{\boldsymbol{\eta}}^z = i\mathbf{c}^+\boldsymbol{\eta}^- - i\mathbf{c}\boldsymbol{\eta}^+ + i\lambda\boldsymbol{\eta}^- - i\lambda\boldsymbol{\eta}^+,$$

instead of the Heisenberg equations

$$\dot{\mathbf{A}} = -i\mathbf{S}^-, \tag{12.17}$$

$$\dot{\mathbf{S}}^- = 2i\mathbf{A}\mathbf{S}^z + 2i\lambda e^{-i\bar{\Delta}\tau}\mathbf{S}^z,$$

$$\dot{\mathbf{S}}^z = i\mathbf{A}^+\mathbf{S}^- - i\mathbf{A}\mathbf{S}^+ + i\lambda e^{i\bar{\Delta}\tau}\mathbf{S}^- - i\lambda e^{-i\bar{\Delta}\tau}\mathbf{S}^+,$$

in the old variables, with time-dependent coefficients. One easily sees that the operator equations (12.16) can be expressed in the form of Heisenberg equations

$$i\frac{d\mathbf{f}}{d\tau} = [\mathbf{f}, \mathbf{H}_1], \tag{12.18}$$

with the following time-independent Hamiltonian

$$\mathbf{H}_1 = N\left\{\mathbf{c}\boldsymbol{\eta}^+ + \mathbf{c}^+\boldsymbol{\eta}^- + \lambda\boldsymbol{\eta}^+ + \lambda\boldsymbol{\eta}^- - \bar{\Delta}\mathbf{c}^+\mathbf{c} - \bar{\Delta}\boldsymbol{\eta}^z\right\}. \tag{12.19}$$

The system (12.16) with Hamiltonian (12.19) possesses in general two global integrals of motion: Hamiltonian (12.19) and also $\mathbf{S}^2$ given by

$$\mathbf{S}^2 = (\boldsymbol{\eta}^z)^2 + \frac{1}{2}(\boldsymbol{\eta}^+\boldsymbol{\eta}^- + \boldsymbol{\eta}^-\boldsymbol{\eta}^+). \tag{12.20}$$

In the case $\lambda = 0$ third integral exists (see (11.45))

$$\mathbf{W} = \mathbf{c}^+\mathbf{c} + \boldsymbol{\eta}^z. \tag{12.21}$$

In this case ($\lambda = 0$) the system (12.16) is completely integrable, and transforms to the quantum Dicke model [24].

Hamiltonian (12.19) does not depend explicitely on time, so it can be treated by the method discussed in section 11. We shall apply this method below to the Hamiltonian (12.19), but before doing so we consider a second, more general way of transforming Hamiltonian (12.13) to remove its time dependence.

B. Effective Time-Independent Hamiltonian.

Here we return to the Hamiltonian (12.13), and apply the method described in sections 3.8 and 8.2. This method replaces (12.13) by a new effective Hamiltonian, which does not depend explicitely on time. First, we introduce new auxiliary boson operators $\mathbf{b}^+$ and $\mathbf{b}$ ($[\mathbf{b}, \mathbf{b}^+] = 1$), and an effective time-independent Hamiltonian

$$\mathbf{H}_2 = \bar{\Delta} N \mathbf{B}^+\mathbf{B} + N[(\mathbf{AS} + \mathbf{A}^+\mathbf{S}) + \varepsilon(\mathbf{B}^+\mathbf{S}^- + \mathbf{BS}^+)], \quad (12.22)$$

where the constant ε will be discussed below, and the operator $\mathbf{B}$ is a normalized boson operator,

$$\mathbf{B} = \mathbf{b}/\sqrt{N}, \quad ([\mathbf{B}, \mathbf{B}^+] = 1/N).$$

We introduce a coherent state at $\tau = 0$ for the operator $\mathbf{B}$ (or $\mathbf{b}$)

$$|\beta> = \exp(N\beta\mathbf{B}^+ - N\beta^*\mathbf{B})|0>, \quad (12.23)$$
$$(\mathbf{B}|\beta> = \beta|\beta>).$$

The average number of photons in this field at $\tau = 0$ is then

$$n_\beta^0 \equiv <\beta|\mathbf{b}^+\mathbf{b}|\beta> = N <\beta|\mathbf{B}^+\mathbf{B}|\beta> = N|\beta|^2.$$

In what follows we assume that $|\beta| \gg 1$, so the field defined by the operators $\mathbf{b}^+$ and $\mathbf{b}$ is "strong". Then, one can see that in the limit

$$\beta = \beta_0 \to \infty, \quad \varepsilon \to 0, \quad \beta_0\varepsilon = \lambda = const., \quad (12.24)$$

the Hamiltonian $\mathbf{H}_2$ (12.22) transforms into (12.13) (for simplicity β_0 is assumed to be real). Since Hamiltonian (12.22) does not depend explicitly on time, we can apply the method used in section 11 with the consequent transition (12.24).

12.4 Closed C-Number Equations for Quantum Expectation Values in Coherent States

Consider first the Hamiltonian $\mathbf{H}_1$ (12.19). Introduce at $\tau = 0$ boson and spin coherent states (11.15) and (11.20), where the operator $\mathbf{J}$ is defined in (11.16). Using the formulae (11.29) produces a c-number differential equation for the time-dependent expectation value $f(\tau)$ (11.25)

$$\frac{\partial f(\tau)}{\partial \tau} = \hat{K}^{(1)} f(\tau), \quad (12.25)$$

where superscript (1) in operator $\hat{K}^{(1)}$ means that this operator is derived for the Hamiltonian $\mathbf{H}^{(1)}$ (12.19). The operator $\hat{K}^{(1)}$ has the form

$$\hat{K}^{(1)} = \hat{K}_{cl}^{(1)} + \frac{1}{N}\hat{K}_q^{(1)}, \tag{12.26}$$

where

$$\hat{K}_{cl}^{(1)} = i[-\bar{\Delta}z^*\frac{\partial}{\partial z^*} + \bar{\Delta}z\frac{\partial}{\partial z} - \bar{\Delta}\xi^*\frac{\partial}{\partial \xi^*} + \bar{\Delta}\xi\frac{\partial}{\partial \xi} \tag{12.27}$$

$$+\frac{2J}{N}\frac{1}{(1+|\xi|^2)}(\xi^*\frac{\partial}{\partial z^*} - \xi\frac{\partial}{\partial z}) + (z^*\frac{\partial}{\partial \xi^*} - z\frac{\partial}{\partial \xi}) + (z^*\xi^2\frac{\partial}{\partial \xi} - z\xi^{*2}\frac{\partial}{\partial \xi^*})$$

$$+\lambda(\frac{\partial}{\partial \xi^*} - \frac{\partial}{\partial \xi} + \xi^2\frac{\partial}{\partial \xi} - \xi^{*2}\frac{\partial}{\partial \xi^*})],$$

and

$$\hat{K}_q^{(1)} = i\left(\xi^2\frac{\partial^2}{\partial z\partial\xi} - \xi^{*2}\frac{\partial^2}{\partial z^*\partial\xi^*}\right). \tag{12.28}$$

We present now the second method of deriving the c-number equation for time-dependent quantum expectation value of arbitrary operator $\mathbf{f}(\tau)$, based on the Hamiltonian $\mathbf{H}_2$ (12.22). Introduce a coherent state at $\tau = 0$

$$|z, \beta, \xi> \equiv |z>|\beta>|\xi>, \tag{12.29}$$

where all coherent states $|z>$, $|\beta>$, and $|\xi>$ are defined in (11.21) and (11.23). Introduce also, analogously to (11.25), the definition for dynamical expectation value of an arbitrary operator $\mathbf{f}(\tau)$ in the state (12.29), namely

$$f(\tau) = <\xi, \beta, z|\mathbf{f}(\tau)|z, \beta, \xi> = f(z^*, z; \beta^*, \beta; \xi^*, \xi; \tau). \tag{12.30}$$

Then, analogously to the scheme used in deriving equation (12.25), we find an equation for $f(\tau)$ for the case of Hamiltonian $\mathbf{H}_2$ (12.22)

$$\frac{\partial f(\tau)}{\partial \tau} = \hat{K}^{(2)} f(\tau). \tag{12.31}$$

The operator $\hat{K}^{(2)}$ may be written as

$$\hat{K}^{(2)} = \hat{K}_0^{(2)} + \hat{K}_\beta^{(2)}, \tag{12.32}$$

where

$$\hat{K}_0^{(2)} = i\{\frac{2J}{N}\frac{1}{(1+|\xi|^2)}(\xi^*\frac{\partial}{\partial z^*} - \xi\frac{\partial}{\partial z}) + (z^*\frac{\partial}{\partial \xi^*} - z\frac{\partial}{\partial \xi}) \tag{12.33}$$

$$+(z^*\xi^2\frac{\partial}{\partial \xi} - z\xi^{*2}\frac{\partial}{\partial \xi^*})$$

$$+\frac{1}{N}\left(\xi^2\frac{\partial^2}{\partial z\partial \xi} - \xi^{*2}\frac{\partial^2}{\partial z^*\partial \xi^*}\right)\},$$

and

$$\hat{K}_\beta^{(2)} = i\bar{\Delta}\left(\beta^*\frac{\partial}{\partial \beta^*} - \beta\frac{\partial}{\partial \beta}\right) + i\frac{\lambda}{\beta_0}\{\frac{2J}{N}\frac{1}{(1+|\xi|^2)}(\xi^*\frac{\partial}{\partial \beta^*} - \xi\frac{\partial}{\partial \beta}) \tag{12.34}$$

$$+(\beta^*\frac{\partial}{\partial \xi^*} - \beta\frac{\partial}{\partial \xi}) + (\beta^*\xi^2\frac{\partial}{\partial \xi} - \beta\xi^{*2}\frac{\partial}{\partial \xi^*})$$

$$-\frac{1}{N}\left(\xi^{*2}\frac{\partial^2}{\partial \beta^*\partial \xi^*} - \xi^2\frac{\partial^2}{\partial \beta\partial \xi}\right)\}.$$

In (12.34) we have used the relation $\beta_0\varepsilon = \lambda$ (see (12.24)). In the limit $\beta \to \infty$ the operator $\hat{K}_\beta^{(2)}$ (12.34) becomes

$$\hat{K}_\beta^{(2)} = i\bar{\Delta}\left(\beta^*\frac{\partial}{\partial \beta^*} - \beta\frac{\partial}{\partial \beta}\right) \tag{12.35}$$

$$+i\frac{\lambda}{\beta_0}\left(\beta^*\frac{\partial}{\partial \xi^*} - \beta\frac{\partial}{\partial \xi} + \beta^*\xi^2\frac{\partial}{\partial \xi} - \beta\xi^{*2}\frac{\partial}{\partial \xi^*}\right).$$

The derivatives with respect to β and β^* may be removed by transforming to a coordinate system rotating with frequency $\bar{\Delta}$

$$\tilde{\beta}^* = e^{i\bar{\Delta}\tau}\beta^*, \quad \tilde{\beta} = e^{-i\bar{\Delta}\tau}\beta. \tag{12.36}$$

In this coordinate system the operator $\hat{K}_\beta^{(2)}$ transforms to

$$\hat{K}_\beta^{(2)} = i\frac{\lambda}{\beta_0}(e^{-i\bar{\Delta}\tau}\tilde{\beta}^*\frac{\partial}{\partial \xi^*} - e^{i\bar{\Delta}\tau}\tilde{\beta}\frac{\partial}{\partial \xi} \tag{12.37}$$

$$+e^{-i\bar{\Delta}\tau}\tilde{\beta}^*\xi^2\frac{\partial}{\partial\xi} - e^{i\bar{\Delta}\tau}\tilde{\beta}\xi^{*2}\frac{\partial}{\partial\xi^*}).$$

In (12.37) the values $\tilde{\beta}^*$ and $\tilde{\beta}$ can be used as parameters. Choosing $\tilde{\beta}^* = \tilde{\beta} = \beta_0$ we derive finally the operator $\hat{K}^{(2)}$

$$\hat{K}^{(2)} = \hat{K}_0 + i\lambda\left(e^{-i\bar{\Delta}\tau}\frac{\partial}{\partial\xi^*} - e^{i\bar{\Delta}\tau}\frac{\partial}{\partial\xi} + e^{-i\bar{\Delta}\tau}\xi^2\frac{\partial}{\partial\xi} - e^{i\bar{\Delta}\tau}\xi^{*2}\frac{\partial}{\partial\xi^*}\right), \tag{12.38}$$

where the operator $\hat{K}_0$ is defined in (12.33).

The explicit dependence on time in the operator (12.38) may be excluded by the substitution

$$\begin{aligned} \tilde{\xi}^* &= \xi^* e^{i\bar{\Delta}\tau}, \quad \tilde{\xi} = \xi e^{-i\bar{\Delta}\tau}, \\ \tilde{z}^* &= z^* e^{i\bar{\Delta}\tau}, \quad \tilde{z} = z e^{-i\bar{\Delta}\tau}. \end{aligned} \tag{12.39}$$

With the substitution (12.39) equation (12.31) takes the form

$$\frac{\partial f(\tau)}{\partial\tau} = \left(\hat{K}_{cl}^{(2)} + \frac{1}{N}\hat{K}_q^{(2)}\right) f(\tau), \tag{12.40}$$

where

$$\begin{aligned} \hat{K}_{cl}^{(2)} = i\{ &-\bar{\Delta}\tilde{z}^*\frac{\partial}{\partial\tilde{z}^*} + \bar{\Delta}\tilde{z}\frac{\partial}{\partial\tilde{z}} - \bar{\Delta}\tilde{\xi}^*\frac{\partial}{\partial\tilde{\xi}^*} \\ &+\bar{\Delta}\tilde{\xi}\frac{\partial}{\partial\tilde{\xi}} + \frac{2J}{N}\frac{1}{(1+|\tilde{\xi}|^2)}\left(\tilde{\xi}^*\frac{\partial}{\partial\tilde{z}^*} - \tilde{\xi}\frac{\partial}{\partial\tilde{z}}\right) \\ &+\left(\tilde{z}^*\frac{\partial}{\partial\tilde{\xi}^*} - \tilde{z}\frac{\partial}{\partial\tilde{\xi}}\right) + \left(\tilde{z}^*\tilde{\xi}^2\frac{\partial}{\partial\tilde{\xi}} - \tilde{z}\tilde{\xi}^{*2}\frac{\partial}{\partial\tilde{\xi}^*}\right) \\ &+\lambda\left(\frac{\partial}{\partial\tilde{\xi}^*} - \frac{\partial}{\partial\tilde{\xi}} + \tilde{\xi}^2\frac{\partial}{\partial\tilde{\xi}} - \tilde{\xi}^{*2}\frac{\partial}{\partial\tilde{\xi}^*}\right), \end{aligned} \tag{12.41}$$

and

$$\hat{K}_q^{(2)} = i\left(\tilde{\xi}^2\frac{\partial^2}{\partial\tilde{z}\partial\tilde{\xi}} - \tilde{\xi}^{*2}\frac{\partial^2}{\partial\tilde{z}^*\partial\tilde{\xi}^*}\right). \tag{12.42}$$

One can see that equations (12.40)-(12.42) coincide with equations (12.25)-(12.28).

This second second approach may be useful in those cases, where there is no simple way to derive a time-independent Hamiltonian. In our case the situation is different, and in what follows we shall use equation (12.25)-(12.28) (without the additional symbol $\sim$). This equation must be solved with initial conditions

$$f(0) = f(z^*, z; \xi^*, \xi).$$

For operator **f** we substitute from the set of the operators

$$\mathbf{A}^+, \quad \mathbf{A}, \quad \mathbf{S}^z, \quad \mathbf{S}^+, \quad \mathbf{S}^-,$$

and, the initial conditions have the form (11.33).

12.5 Semiclassical Approximation and Chaotic Dynamics

The semiclassical approximation corresponds to the limit (11.34). In this case the small quantum parameter goes to zero

$$\frac{1}{N} = \frac{\hbar}{I_{field}} \to 0, \quad (I_{fiefd} = \hbar N), \tag{12.43}$$

where I_{field} is the action possessed by the field in the resonator, in the process of dynamical interaction with the atomic sub-system. The limit (11.34) corresponds to considering of the radiation field sub-system as classical. In this case the operator $\hat{K}_q^{(1)}$ (12.28) does not contribute to the dynamics of the expectation values, and only the operator $\hat{K}_{cl}^{(1)}$ needs to be taken into account. The following equation then defines the slow dynamics of our system

$$\frac{\partial f(\tau)}{\partial \tau} = \hat{K}_{cl}^{(1)} f(\tau), \quad f(0) = f(z^*, z; \xi^*, \xi). \tag{12.44}$$

Since the differential operator $\hat{K}_{cl}^{(1)}$ contains only first order derivatives, a solution of (12.44) may be found by the method of characteristics. These semiclassical equations for the characteristics are,

$$\frac{dz}{d\tau} = i\bar{\Delta} z - i\frac{g\xi}{(1+|\xi|^2)}, \quad g = \frac{2J}{N}, \tag{12.45}$$

$$\frac{d\xi}{d\tau} = i\bar{\Delta}\xi - iz + iz^*\xi^2 - i\lambda + i\lambda\xi^2.$$

An analysis of this semiclassical dynamics (presented in a form different from (12.45)) is given in [102]. In [102] a transition from regular dynamics to global (strong) chaos is found in this system under the approximate conditions (for $g = 2J/N = 1$)

$$\lambda \sim |\bar{\Delta}| \sim 1. \tag{12.46}$$

Condition (12.46) means that the characteristic "slow" dimensional frequences in this system: the cooperative frequency ω_c (12.4), the duration frequency Δ (12.8), and the Rabi frequency $\omega_R \sim 2\gamma/\hbar$, connected with the external field in (12.2), all have the same order

$$\omega_c \sim \omega_R \sim |\Delta|. \tag{12.47}$$

Conditions (12.46), (12.47) correspond to a rather strong resonant interaction in the system which produces global chaos. In [103,104] weak semiclassical chaos is analysed in detail under the condition $\lambda \ll 1$. In this case the Smale-horseshoe mechanism is responsible for producing chaotic behavior. Conditions (12.46), (12.47) often occur in experiments (see, for example, [25]). We need not describe the semiclassical dynamics in detail here only we present a few typical results, which will be used later when comparing with quantum-mechanical considerations. First, note that the system of equations (12.45) is expressible in Hamiltonian form with the Hamiltonian

$$H_{cl} = \bar{\Delta}(q^2 + p^2 - x^2 - y^2) + 2(g - q^2 - p^2)^{1/2} \cdot (xq - yp + \lambda q) = E, \tag{12.48}$$

upon introducing the following new real variables and rescalling time

$$z = x + iy, \quad \zeta \equiv \frac{\sqrt{g}|\xi|}{\xi\sqrt{1 + |\xi|^2}} = q + ip, \tag{12.49}$$

$$\tau' = 2\tau, \quad g = 2J/N.$$

Consider first the case $\lambda = 0$, where three integrals of motion exist: $\mathbf{H}_1$ (12.19), $\mathbf{S}^2$ (12.20), and $\mathbf{W}$ (12.21). In the semiclassical limit for $\lambda = 0$ the integral W has the form

$$W = \langle \mathbf{W} \rangle = x^2 + y^2 - q^2 - p^2. \tag{12.50}$$

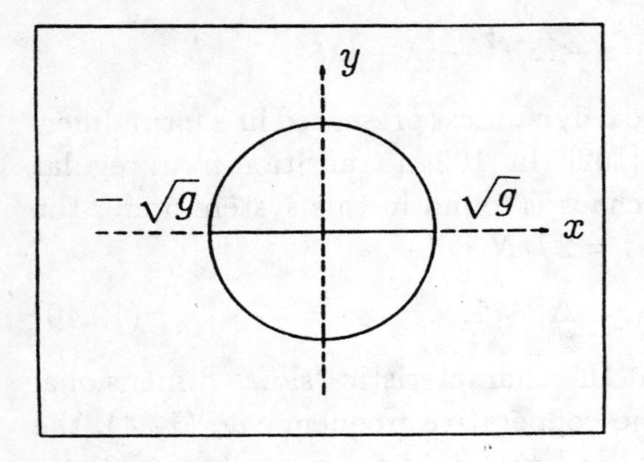

Fig. 29. The separatrix for semiclassical integrable motion, solid line, $\lambda = 0$.

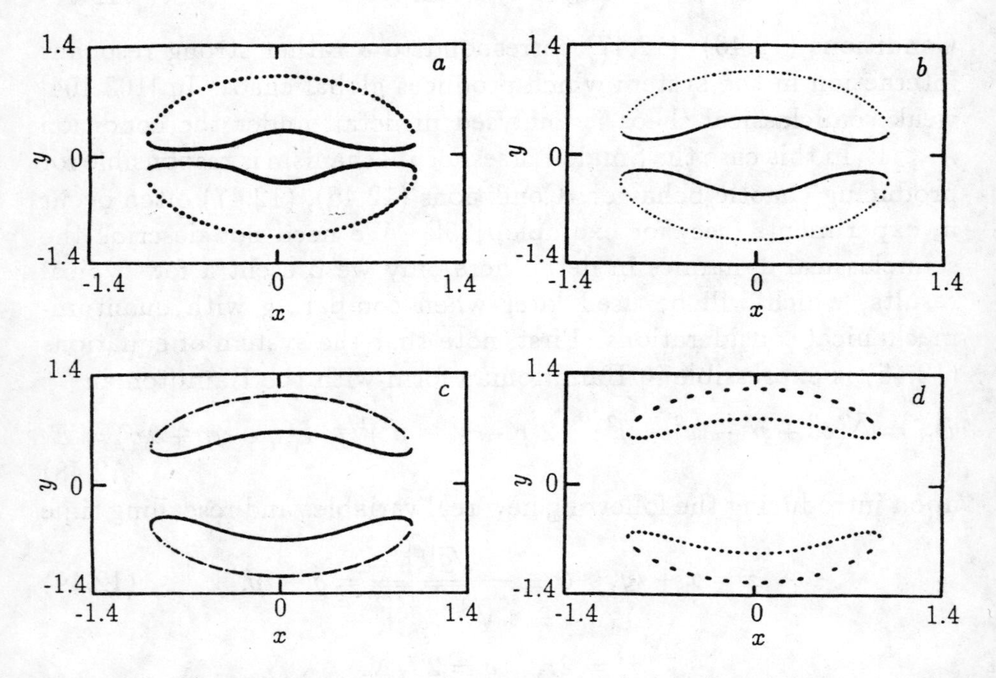

Fig. 30. The Poincare map for one semiclassical trajectory on the plane (x, y) when this trajectory crosses the plane $q = 0$. Integrable case ($\lambda = 0$); $\bar{\Delta} = 1$; dependence on the value of integral W;(a) W=0.1; (b) W=0.2; (c) W=0.4; (d) W=0.6; $E = 0; g = 1$.

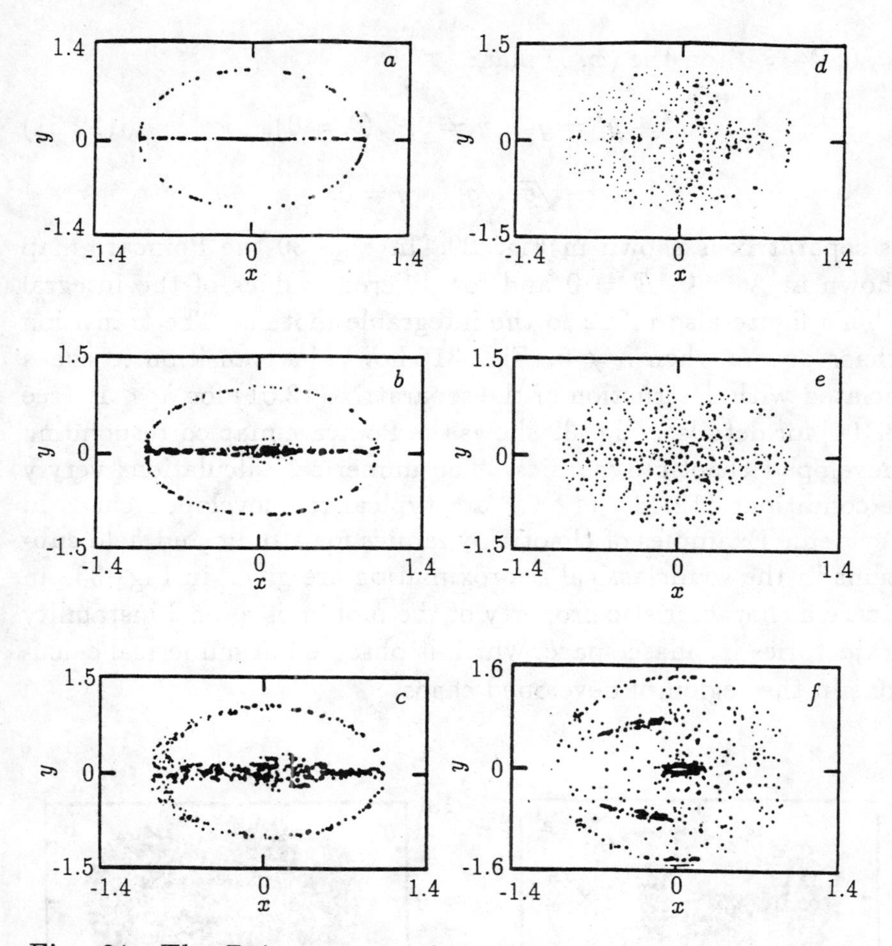

Fig. 31. The Poincare map for one semiclassical trajectory on the plane (x, y) when this trajectory crosses the plane $q = 0$. Dependence on the interaction constant λ; $\bar{\Delta} = 1$; $(E = 0; g = 1)$. Initial conditions: $W(0) = 0; p(0) = 0; q(0) = 0$; (a) integrable case, trajectory close to separatrix $(\lambda = 0)$. Nonintegrable case: (b), (c) destaction of the separatrix; (b) $\lambda = 0.01$; (c) $\lambda = 0.03$; (d)-(f) transition to developed chaos; (d) $\lambda = 0.07$; (e) $\lambda = 0.1$; (f) $\lambda = 0.5$.

We construct a Poincare map of the (x, y) plane from each crossing by the trajectory of the plane $q = q_0 = const.$, with $p > 0$. Choosing $q_0 = 0$ gives the following equations for the separatrix

$(E = 0; W = 0)$ on the (x, y) plane

$$x^2 + y^2 = g, \quad y \neq 0, \quad (\lambda = 0), \tag{12.51}$$

$$x \epsilon [-\sqrt{g}, \sqrt{g}], \quad y = 0.$$

This separatrix is shown in Fig. 29. In Fig. 30 the Poincare map is shown at $\lambda = 0$, $E = 0$ and for different values of the integral W. This figure also refers to the integrable motion. The transition to chaos occurs when $\lambda \neq 0$. Fig. 31 shows the transition to chaos associated with destruction of the separatrix (12.51) for $\lambda \ll 1$. (see [103,104] for details.) Fig. 32 shows the Poincare map corresponding to developed chaotic dynamics. The numerical calculations veryfy that conditions (12.46), (12.47) are typical for developed chaos in the system. Examples of chaotic dynamics for atomic and field subsystems in the semiclassical approximation are given in Fig. 33. In this case a characteristic property of the motion is a local instability of trajectories in phase space, which is observed in numerical calculations in the region of developed chaos.

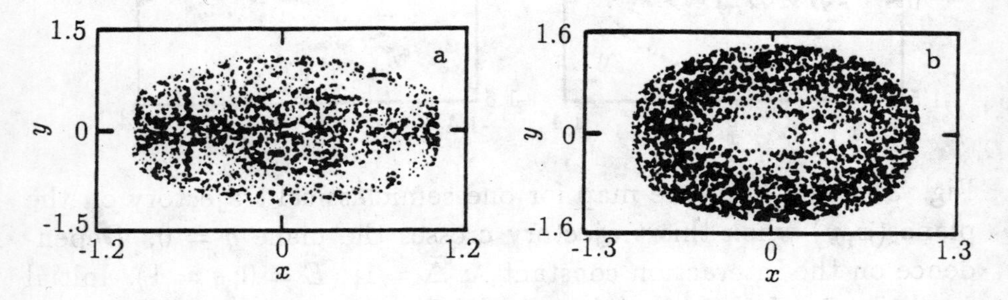

Fig. 32. Poincare map for semiclassical motion; the case of developed chaos (long time simulation: $\tau_{max} = 500$): $g = 1; z(0) \approx (0.65; -0.35); \xi(0) \approx (0; -1.73)$; (a) $\lambda = 0.1$; (b) $\lambda = 2$.

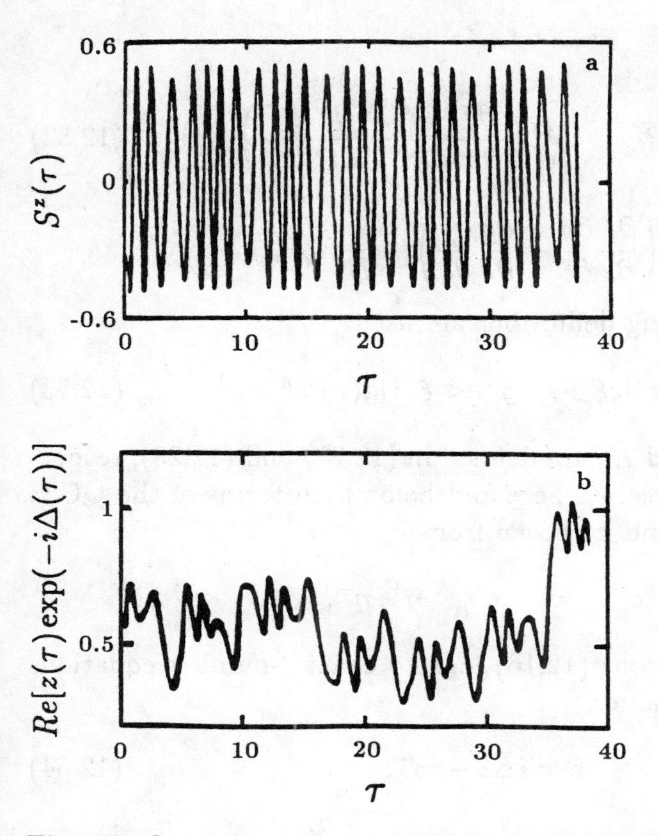

Fig. 33. Semiclassical chaotic dynamics of $S^z(\tau)$- (a); and $Re[z(\tau)\exp(-i\bar{\Delta}\tau)]$-(b); $\bar{\Delta} = 1; g = 1; \lambda = 2; z(0) \approx (0.65; -0.34)$; $\xi(0) \approx (0; -1.75)$.

12.6 The Peculiarities of Quantum Dynamics in the Regime of Strong Semiclassical Chaos

Together with consideration of the dynamics of quantum expectation values, we also investigated the dynamics of QCFs of two arbitrary operators **f** and **g** (11.35). The exact equation for QCF $P_{f,g}(\tau)$

(11.35) has the form

$$\frac{\partial P_{f,g}}{\partial \tau} = \hat{K}_{cl} P_{f,g} - i\xi^{*2} \left(\frac{\partial f}{\partial z^*} \frac{\partial g}{\partial \xi^*} + \frac{\partial g}{\partial z^*} \frac{\partial f}{\partial \xi^*} \right) \tag{12.52}$$

$$+ i\xi^2 \left(\frac{\partial f}{\partial z} \frac{\partial g}{\partial \xi} + \frac{\partial g}{\partial z} \frac{\partial f}{\partial \xi} \right) + \frac{1}{N} \hat{K}_q P_{f,g}.$$

In (12.52) the following definitions are used

$$f \equiv < \xi, z | \mathbf{f}(\tau) | z\xi >, \quad g \equiv < \xi, z | \mathbf{g}(\tau) | z\xi >, \tag{12.53}$$

and operators $\hat{K}_{cl}$ and $\hat{K}_q$ are defined in (12.27) and (12.28), respectively (we omit superscip 1 here and below). In terms of the QCFs $P_{f,g}$ for operators $\mathbf{f}$ and $\mathbf{g}$ chosen from

$$\mathbf{c}^+, \quad \mathbf{c}, \quad \boldsymbol{\eta}^z, \quad \boldsymbol{\eta}^+, \boldsymbol{\eta}^-,$$

the Heisenberg equations (12.16) lead to exact c-number equations for the expectation values

$$\dot{c} = i\bar{\Delta} c - i\eta^-, \tag{12.54}$$

$$\dot{\eta}^- = i\bar{\Delta}\eta^- + 2i\lambda\eta^z + 2i\eta^z c + \frac{2i}{N} P_{\eta^z, c},$$

$$\dot{\eta}^z = ic^+\eta^- - ic\eta^+ + i\lambda\eta^- - i\lambda\eta^+ + \frac{i}{N}(P_{c^+,\eta^-} - C.C.),$$

where $c^+, c, \eta^z, \eta^+, \eta^-$ are c-number functions. We also present expressions for the expectation values of the integrals of motion $\mathbf{H}_1$ (12.19), $\mathbf{S}^2$ (12.20), and $\mathbf{W}$ (12.21) (the last of which is only an integral of motion when $\lambda = 0$)

$$E_q \equiv < \mathbf{H}_1 > = c\eta^+ + c^+\eta^- + \lambda\eta^+ + \lambda\eta^- \tag{12.55}$$

$$-\bar{\Delta} c^+ c - \bar{\Delta}\eta^z + \frac{1}{N}(P_{c,\eta^+} + C.C.) - \frac{1}{N}\bar{\Delta} P_{c^+,c},$$

$$< \mathbf{S}^2 > = \frac{J(J+1)}{N^2} = (\eta^z)^2 + \eta^+\eta^- + \frac{1}{N} P_{\eta^z,\eta^z} + \frac{1}{2N}(P_{\eta^+,\eta^-} + C.C.),$$

$$W \equiv \langle \mathbf{W} \rangle = \eta^z + c^+ c + \frac{1}{N} P_{c^+, c}.$$

The system of equations (12.52), (12.54) for quantum expectation values and QCFs is exact and closed. The QCFs $P_{f,g}$ (11.35) vanish in the semiclassical limit (11.34): i.e., $P^{sc}_{f,g}(\tau) = 0$ for arbitrary τ.

We use an approximate scheme for numerical integration of equations (12.52), (12.54) analogous to that described in section 11. Since the quantum parameter ϵ is small: $\epsilon = 1/N \ll 1$ (in our numerical experiments the number of atoms reached the value $N_{max} \sim 10^9$), we first integrate the equations (12.54) without taking into account the QCFs $P_{f,g}(\tau)$. Then, dynamical expectation values $f(\tau)$ and $g(\tau)$ are substituted into equations (12.53), and QCFs $P_{f,g}(\tau)$ are found numerically. When calculating QCFs $P_{f,g}(\tau)$ the term $\sim 1/N$ in (12.53) is neglected at this order. After this, the QCFs $P_{f,g}(\tau)$ are substituted into equation (12.54), and quantum expectation values $f(\tau)$ and $g(\tau)$ are numerically derived.

We present the results of numerical calculations. The time-scale $\tau_\hbar(N)$ at which semiclassical approximation is violated is calculated numerically for both regular and chaotic semiclassical motion. This time-scale is calculated analogously to the scheme described in section 11. For this we consider a vector

$$\vec{D}(\tau) = \left(c(\tau), \eta^+(\tau), \eta^z(\tau), \right), \tag{12.56}$$

which characterizes the dynamics of quantum expectation values ("quantum trajectory"). Correspondingly, in the semiclassical limit (11.34) the function $\vec{D}(\tau)$ (12.56) reduces to $\vec{D}_{cl}(\tau)$, which is calculated using the semiclassical equations of motion. Then, analogously to (11.48), we introduce a function $\delta(\tau)$

$$\delta(\tau) = \frac{|c(\tau) - c_{cl}(\tau)| + |\eta^+(\tau) - \eta^+_{cl}(\tau)| + |\eta^z(\tau) - \eta^z_{cl}(\tau)|}{|c_{cl}(\tau)| + |\eta^+_{cl}(\tau)| + |\eta^z_{cl}(\tau)|}, \tag{12.57}$$

which characterizes the size of the quantum corrections relative to the semiclassical dynamics, provided $\delta(0)$ is small (in the numerical calculations presented below $\delta(0) = 0$). The time-scale $\tau_\hbar$ is calculated as the solution of the equation $\delta(\tau_\hbar) = const..$ for a particular choice of N, the number of atoms. We have chosen *const.* = 0.01 (or 0.02). The number of atoms N is varied according to the law: $N = 2^n, (n = 1, 2, ..., 30)$, so N_{max} reaches the value $N_{max} \sim 10^9$.

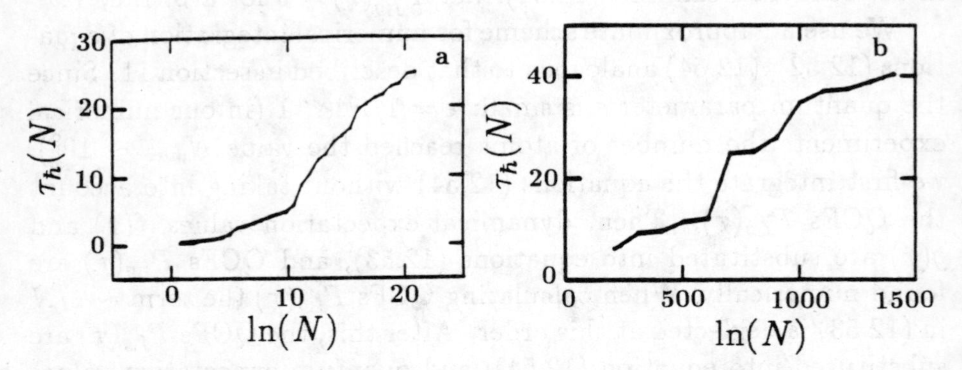

Fig. 34. Dependence $\tau_\hbar(N)$; (a)-for the case of developed chaos in the semiclassical limit: $\lambda = 0.5$; z(0)=(0.65,-0.35); $\xi(0) = (0, -1.73)$; (b)- for regular semiclassical dynamics: $\lambda = 0.1$; $\bar{\Delta} = 1$; z(0)=0; $\xi(0) \approx (0.33; 0)$; a 1% criterion was used.

Simultaneously, the integrals E_q and W are calculated. In Figs. 34a,b the characteristic dependence $\tau_\hbar(N)$ is presented for the case of developed semiclassical chaos (Fig. 34a), and for the case of regular motion in the semiclassical limit (Fig. 34b). Fig. 34a shows for the case of developed semiclassical chaos, that the dependence $\tau_\hbar(N)$ is close to (11.52), i.e., nearly logarithmic in N. For regular semiclassical dynamics the time-scale $\tau_\hbar(N)$ significantly increases (compare Fig. 34a and Fig. 34b). Figs. 35a,b show the results of numerical calculation of the QCF $P(\tau)$, defined by

$$P(\tau) = |P_{c^+,c}(\tau)| + |P_{s^z,c}(\tau)| + |P_{c^+,s^-}(\tau)|. \tag{12.58}$$

Fig. 35a corresponds to the chaotic semiclassical motion, and Fig. 35b corresponds to the regular semiclassical motion. It follows from Fig. 35a that under conditions of developed semiclassical chaos the QCF $P(\tau)$ grows in time nearly exponentially.

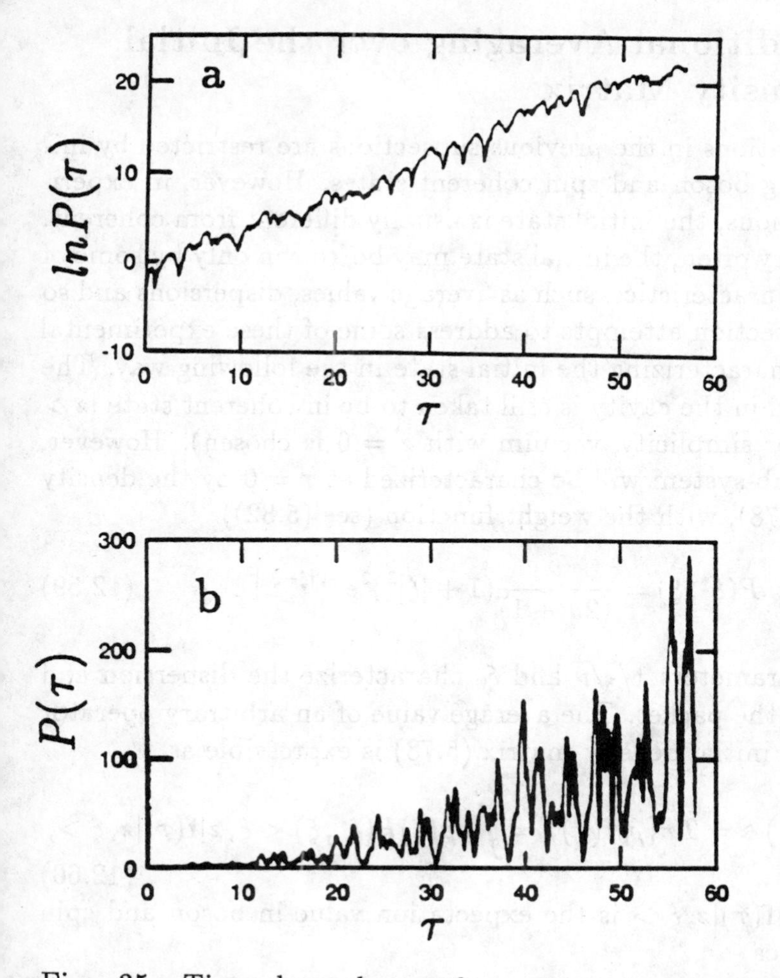

Fig. 35. Time dependence of quantum correlation function $P(\tau)$ in coherent state, for the case of developed semiclassical chaos: (a) $\bar{\Delta} = 1; \lambda = -1; g = 1; z(0) = (0,0); \xi(0) = (1,0)$; and for the case of regular semiclassical dynamics: (b) $\bar{\Delta} = 3; \lambda = 1; g = 1; z(0) = (0,0); \xi(0) = (1,0)$.

When the dynamics in the semiclassical limit is regular the QCF $P(\tau)$ grows approximately according to a power law (see Fig. 35b).

12.7 Additional Averaging over the Initial Density Matrix

The considerations in the previous subsections are restricted by initially choosing boson and spin coherent states. However, in experimental situations, the initial state is usually different from coherent. Moreover, very often, the initial state may be known only by some of its integral characteristics, such as average values, dispersions and so on. This subsection attempts to address some of these experimental features by characterizing the initial state in the following way. The radiation field in the cavity is still taken to be in coherent state $|z>$ at $\tau = 0$ (for simplicity, vacuum with $z = 0$ is chosen). However, the atomic sub-system will be characterized at $\tau = 0$ by the density matrix $\hat{\rho}_0$ (5.78), with the weight function (see (5.82))

$$P(\xi^*, \xi) = \frac{\nu}{(2J+1)}(1 + |\xi|^2)^2 e^{-\nu|\xi-\xi_0|^2}, \qquad (12.59)$$

where the parameters $1/\sqrt{\nu}$ and ξ_0 characterize the dispersion and the center of the packet. The average value of an arbitrary operator $\mathbf{f}(\tau)$ over the initial density matrix (5.78) is expressible as

$$\bar{f}(\tau) \equiv < \mathbf{f}(\tau) > = Tr\{\hat{\rho}_0 \mathbf{f}(\tau)\} = \int d\mu(\xi) P(\xi^*, \xi) < \xi, z|\mathbf{f}(\tau)|z, \xi >, \qquad (12.60)$$

where $< \xi, z|\mathbf{f}(\tau)|z, \xi >$ is the expectation value in boson and spin coherent states

$$< \xi, z|\mathbf{f}(\tau)|z, \xi > \equiv f(z^*, z; \xi^*, \xi; \tau). \qquad (12.61)$$

Using the weight function $P(\xi^*, \xi)$ in the form of a Gaussian-type packet (12.59), we derive a simple formula for $\bar{f}(\tau)$

$$\bar{f}(\tau) = \frac{\nu}{\pi} \int d^2\xi e^{-\nu|\xi-\xi_0|^2} f(z^*, z; \xi^*, \xi; \tau). \qquad (12.62)$$

In the numerical calculations the weight function (12.59) is used. In Figs. 36a,b we present the dependence $\tau_\hbar(N)$ for the case of strong semiclassical chaos.

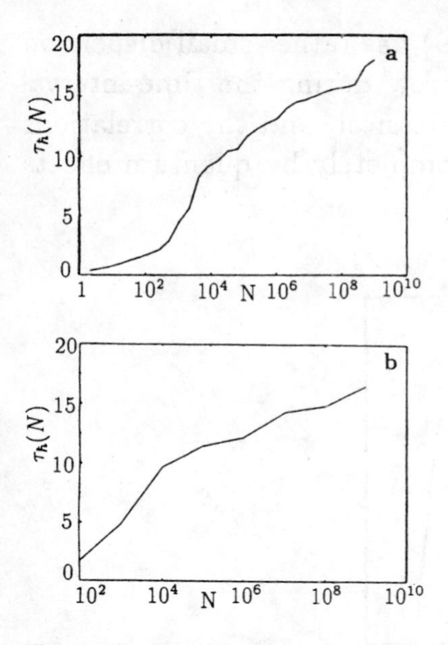

Fig. 36. Dependence $\tau_\hbar(N)$ in the logarithmic scale; $\bar{\Delta} = 1$; $\lambda = -1$; (a) the coherent state at $\tau = 0$ with $z(0) = (0,0)$; $\xi(0) = (0.5, 0)$; (b) the additional averaging over the initial packet (7.3) with $1/\sqrt{\nu} = 0.3$; $z(0) = (0,0)$; the coordinates of the center of the packet: $\xi_0 = (0.5, 0)$. The number of trajectories which were used, $M = 100$.

Fig. 36a corresponds to preparing the initial population of the system in a coherent state, and Fig. 36b represents the same function, but with an additional averaging over the wide initial packet (12.59). In this case, the classical distribution function was chosen in accordance with (12.59). Notice that the dependence $\tau_\hbar(N)$ in both cases is actually the same, and is close to the logarithmical law (1.2). We also compute the time-dependence of QCFs (11.35) taking into account of the additional averaging with the weight function (12.59), for example,

$$\frac{\bar{P}_{c^+,c}(\tau)}{N} = Tr\{\hat{\rho}\mathbf{c}^+(\tau)\mathbf{c}(\tau)\} - Tr\{\hat{\rho}\mathbf{c}^+(\tau)\}Tr\{\hat{\rho}\mathbf{c}(\tau)\}. \qquad (12.63)$$

Note that the averaged QCF's defined this way have both classical and quantum contributions, and do not vanish in the semiclassical limit. The result of these calculations may be formulated in the following way. If the initial packet (12.59) has rather small dispersion ($\nu > N$), the quantum correlations grow during the time-interval $0 < \tau < \tau_\hbar$ significantly faster than classical, and the correlations (12.63) are determined in this case completely by quantum effects (see Fig. 37).

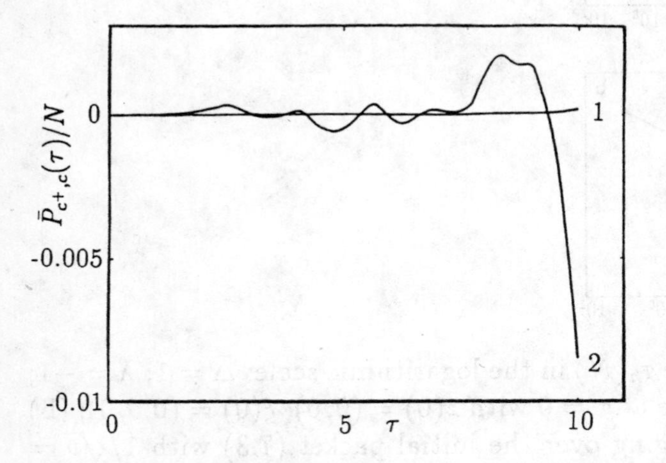

Fig. 37. Time dependence of the field correlation function (7.5); $\bar{\Delta} = 1$; $\lambda = -1$; $N = 10^4$; $1/\sqrt{\nu} = 0.01$; $z(0) = (0,0)$. The coordinates of the center of the packet: $\xi_0 = (0.5, 0)$. The curve 1 corresponds to the classical correlation function; the curve 2 corresponds to the quantum correlation function.

This effect could be observed experimentally. In the case of a rather wide initial packet the effect of quantum correlations is difficult to separate from the growing classical correlations. Numerical calculations show that this time-behavior is typical for various QCFs $\bar{P}_{f,g}(\tau)$. Note also, that in all cases considered above the integrals E_q and W are conserved for times $0 < \tau < \tau_\hbar$ with approximately 0.2% accuracy.

12.8 Conclusion

In section 12 we consider the system of N two-level atoms in a resonator with rather large quality factor, Q, interacting resonantly with a single eigenmode of the field, and with an external coherent field with a frequency slightly different from a frequency of atomic transition. The main problem of interest in this section is to estimate the role of quantum effects under conditions of developed chaos in the semiclassical approximation (when the radiation field is considered classically). Developed semiclassical chaos appears in this system at rather usual experimental conditions

$$\omega_c \sim \omega_R \sim |\Delta|,$$

where ω_c is the cooperative frequency (which includes $\sqrt{N}$ dependence); ω_R is the Rabi frequency of an individual two-level atom in the resonant external field, and Δ is the difference between the frequencies of the external field and the atomic transition.

Under conditions of developed chaos in the semiclassical limit, numerical calculations show that the dynamics of quantum expectation values begins differing significantly from the corresponding semiclassical solutions on logarithmically small times: $\tau_\hbar(N) \sim C_1 \ln(C_2 N)$, where the number of atoms N plays the role of the quasiclassical parameter

$$N \sim \frac{I_{field}}{\hbar},$$

where I_{field} is a characteristic amplitude of the action of the radiation field. We would like to emphasize here that even at apparently large values of N (in computer experiment N reached the value $N_{max} \sim 10^9$), quantum effects become detectable after rather small times. For example, according to the data presented in Fig. 34a, for $\ln N \approx 10$ ($N \approx 22026$), quantum effects at the level of 1% can be observed already at times $\tau = \omega_c t \approx 5$. Quantum correlation functions grow in this regime exponentially during the same time interval $0 < \tau < \tau_\hbar \sim \ln N$. This predicted QCF growth might admit experimental verification. The results presented in this section show that the effects discussed could be observed experimentally, for example, in a system of two-level Rydberg atoms in a microwave cavity with parameters

$$Q/\omega \gg \omega_c, \omega_R, \Delta$$

which are experimental conditions close to those realized in [26,27] (see also review [25]).

The system (12.52)-(12.54) considered in this paper provides a useful means of studying quantum chaos phenomena in the quasiclassical region ($N \gg 1$), when the atom-field interaction dominates the dynamics, and the influence of relaxation processes is small during the (finite) times of interest.

13 Violation of the Semiclassical Approximation and Quantum Chaos in a Paramagnetic Spin System

13.1 Introduction

In this section we consider quantum chaos in a magnetic system whose classical/quantum transition and associated "$\ln N$- effect" may turn out to be experimentally observable. The system consists of N paramagnetic spins ($s = 1/2$) which are placed in a resonator with large quality factor, Q. These spins interact with an external homogeneous magnetic field B_0. We assume there is no direct interaction between these spins. Their interaction is solely through a single electromagnetic eigenmode of the resonator, and this mode is chosen to be in resonance with the frequency of spin precession in the external magnetic field B_0. In this part of the problem formulation, our system is analogous to the quantum Dicke model [24]. We then add to our spin system a periodically-modulated magnetic field, perpendicular to B_0, with constant amplitude and with modulation frequency slightly different from the eigenmode frequency (or the spin precession frequency). This field acts on the spins as a nearly-resonant periodic external force. We show that the Hamiltonian which describes a slow dynamics in this magnetic system can be reduced to the Hamiltonian $\mathbf{H}_1$ (12.13) which describes slow dynamics in the nonlinear optical system considered in the previous section. Three characteristic frequences appear in the slow dynamics of our magnetic system: (1) the cooperative Dicke frequency ω_c which describes slow self-consistent energy exchange between the spins and the resonator eigenmode; (2) the Rabi frequency ω_R which is proportional to the amplitude of the external coherent magnetic field and describes slow nutation of an individual spin in resonance with electromagnetic field; and (3) the difference Δ between the frequency of spin precession in magnetic field B_0 and the frequency of the external coherent magnetic field. In the semiclassical approximation (when the radiation field in the resonator is considered as a classical subsystem) this paramagnetic spin system shows "developed" (global) chaos under conditions that may be expressed roughly as follows: $\omega_c \sim \omega_R \sim |\Delta|$. As before, this condition means that in slow dynamics the different nonlinear resonances strongly interact. We take into account quantum effects by treating the eigenmode of the resonator as a quantum linear oscillator. In this quantum system, the self-consistent field (eigenmode) action is big: $I_{field} \sim \hbar N \gg \hbar$ and the system is in the "deep" quasiclassical region.

We show that for this spin system in the parameter region where developed chaos occurs in the semiclassical approximation, quantum effects lead (as in the nonlinear optical system considered in section 12), to violation of the semiclassical description on the time-scale $\tau_\hbar \sim \ln N$. Also, in this case quantum correlation functions qrow exponentially in the time before $\tau_\hbar$, at a rate that may be experimentally measurable. This section is organized as follows. Section 13.2 describes the parametric spin model. In section 13.3 the Hamiltonian for the slow dynamics of the spin model is derived, and it is shown that it coincides with the Hamiltonian (12.13) considered in section 12. In the Conclusion, we discuss the experimental measurability of "$\ln N$-effect" in this system.

13.2 Description of the Paramagnetic Spin Model

We are dealing with an ensemble of N paramagnetic atoms, each with spin one-half, placed in a cavity resonator and interacting self-consistently via a single resonator eigenmode. These spins also interact with an externally-imposed, homogeneous, magnetic field with two orthogonal components, B_0 and $b(t)$. The first of these, B_0, is constant and has its spin presession frequency tuned to the frequency of the resonator eigenmode. The other homogeneous external magnetic field component, $b(t)$, is modulated periodically in time, with modulation frequency slightly detuned from that of the resonator eigenmode (see Fig. 38).

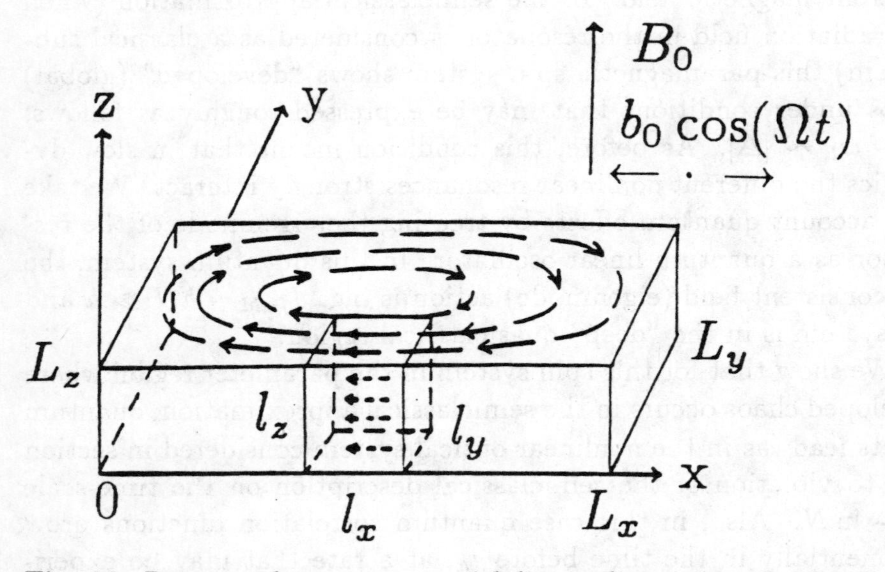

Fig. 38. Rectangular resonator with linear dimensions L_x, L_y, L_z and a paramagnetic sample with dimensions $l_x \ll L_x; l_y \ll L_y; l_z = L_z$. A constant magnetic field b_0 is directed along the z axis; periodic magnetic field $b_0 \cos(\Omega\tau)$ is linearly polarized along the x axis. The curves show the distribution of magnetic field in a chosen resonator eigenmode (this field is homogeneous in the z direction).

We assume we can excite in the cavity resonator a single eigenmode, characterized by its eigenfrequency $\omega = 2\pi f$ and by the geometry of its electric and magnetic field lines. Thus, the amplitudes of all other modes are taken to be negligibly small in comparison with the amplitude of the excited mode. The frequency of magnetic resonance is determined by the constant external magnetic field, B_0, and may be tuned to match the frequency of the excited eigenmode f.

For example, in electron paramagnetic resonance (PMR) assuming $B_0 = 1$ kOe and $g = 2$ gives

$$f = \frac{g\mu_B B_0}{2\pi\hbar} \sim 3GHz.$$

The amplitude of the time-periodic external magnetic field, $b(t)$, may be varied over a wide range: typical values are $b \sim 0.01$ to 10^2 Oe. A typical cavity quality factor is $Q \sim 10^4$, or even higher [25]. (Here $Q = \omega/2\gamma$ and γ is the damping coefficient of our eigenmode.) The dynamics may be taken to be nondissipative over times t less than order $10^4/\omega$. The possibility of making experimental measurements of effects connected with time-scale $\tau_\hbar$ less than this will be discussed below.

The eigenmodes $\vec{E}_{\vec{k}}(\vec{r})$ in the ideal cavity resonator are described by the Helmholtz equation,

$$\Delta \vec{E}_{\vec{k}} + \left(\frac{\omega_{\vec{k}}}{c}\right)^2 \vec{E}_{\vec{k}} = 0, \quad \vec{\nabla} \cdot \vec{E}_{\vec{k}} = 0, \tag{13.1}$$

with tangential boundary conditions $\vec{E}_{\vec{k}} \times \vec{n} = 0$ (where $\vec{n}$ is a normal vector to the surface of resonator). The magnetic field $\vec{B}_{\vec{k}}(\vec{r})$ also satisfies equation (13.1) with normal boundary conditions, $\vec{B}_{\vec{k}} \cdot \vec{n} = 0$. The eigenfrequencies depend on the shape of the resonator, and for a rectangular parallelepiped with sides L_x, L_y, L_z, one has

$$\left(\frac{\omega_{\vec{k}}}{c}\right)^2 = k_x^2 + k_y^2 + k_z^2, \tag{13.2}$$

where

$$k_{x,y,z} = \frac{n_{x,y,z}}{L_{x,y,z}},$$

and $n_{x,y,z} = 0, 1, \ldots$.

For quantum consideration, as is well known, the electromagnetic field in the resonator may be represented in the form

$$\vec{E}(\vec{r}, t) = \sum_{\vec{k}} \mathbf{p}_{\vec{k}}(t) \vec{E}_{\vec{k}}(\vec{r}), \tag{13.3}$$

$$\vec{B}(\vec{r},t) = -\sum_{\vec{k}} \omega_{\vec{k}} \mathbf{q}_{\vec{k}}(t) \vec{B}_{\vec{k}}(\vec{r}),$$

where $\mathbf{p}_{\vec{k}}(t)$ and $\mathbf{q}_{\vec{k}}(t)$ are impulse and coordinate operators, with commutation relations $[\mathbf{q}_{\vec{k}}, \mathbf{p}_{\vec{k}'}] = i\hbar\delta_{\vec{k},\vec{k}'}$. These operators $\mathbf{q}_{\vec{k}}$ and $\mathbf{p}_{\vec{k}}$ are expressible in terms of creation and annihilation operators $\mathbf{a}^+$ and $\mathbf{a}$, as

$$\mathbf{q}_{\vec{k}} = \left(\frac{\hbar}{2\omega_{\vec{k}}}\right)^{1/2} (\mathbf{a}^+_{\vec{k}} + \mathbf{a}_{\vec{k}}), \tag{13.4}$$

$$\mathbf{p}_{\vec{k}} = i\left(\frac{\hbar\omega_{\vec{k}}}{2}\right)^{1/2} (\mathbf{a}^+_{\vec{k}} - \mathbf{a}_{\vec{k}}), \quad [\mathbf{a}_{\vec{k}}, \mathbf{a}^+_{\vec{k}'}] = \delta_{\vec{k},\vec{k}'}.$$

Hence, we arrive at the standard formula for the Hamiltonian of an electromagnetic field in a resonator,

$$\mathbf{H}_f = \frac{1}{8\pi}\int(\vec{B}^2 + \vec{E}^2) = \sum_{\vec{k}} \hbar\omega_{\vec{k}} \left(\mathbf{a}^+_{\vec{k}}\mathbf{a}_{\vec{k}} + \frac{1}{2}\right). \tag{13.5}$$

The interaction of spins in a sample placed in this resonator with a magnetic field is governed by the Hamiltonian

$$\mathbf{H}_{int} = g\mu_B \left(\vec{B}(\vec{r}_0,t) + \vec{B}_{ex}\right) \cdot \sum_{j=1}^{N} \vec{\mathbf{S}}_j. \tag{13.6}$$

In (13.6) $\vec{B}_{ex}$ is an external magnetic field that is assumed to be homogeneous over the sample's size and could depend on time; $\vec{\mathbf{S}}_j$ are spin operators; N is the number of spins in the sample. We have substituted $\vec{B}(\vec{r}_0,t)$ for $\vec{B}(\vec{r},t)$ in (13.6), with

$$\vec{B}(\vec{r}_0,t) = -\sum_{\vec{k}} \left(\frac{\hbar\omega_{\vec{k}}}{2}\right)^{1/2} \vec{B}_{\vec{k}}(\vec{r}_0)(\mathbf{a}^+_{\vec{k}} + \mathbf{a}_{\vec{k}}), \tag{13.7}$$

since the size of the sample being placed into the resonator near location $\vec{r} \approx \vec{r}_0$ is supposed to be small in comparison with the characteristic inhomogeneity of the chosen resonator eigenmode. The magnetic field in such a resonator may be regarded as homogeneous.

In what follows, we shall limit ourselves to the single-mode approximation and consider the following eigenmode of the resonator

$$E_z(x,y) = \sqrt{\frac{16\pi}{V}} \sin(k_x x) \sin(k_y y), \quad E_x = E_y = 0, \tag{13.8}$$

$$B_x(x,y) = -\frac{ck_y}{\omega}\sqrt{\frac{16\pi}{V}} \sin(k_x x) \cos(k_y y),$$

$$B_y(x,y) = \frac{ck_x}{\omega}\sqrt{\frac{16\pi}{V}} \cos(k_x x) \sin(k_y y),$$

$$B_z = 0,$$

where

$$k_{x,y} = \frac{\pi}{L_{x,y}}, \quad V = L_x L_y L_z, \quad \omega^2 = c^2(k_x^2 + k_y^2). \tag{13.9}$$

The normalization of the mode is chosen as

$$\int \vec{E}^2 dV = \int \vec{B}^2 dV = 4\pi.$$

According to (13.8) the electromagnetic field is homogeneous along the z axis. Thus, it is sufficient to assume that the dimensions of the sample along x and y axes are small

$$l_x \ll L_x, \quad l_y \ll L_y. \tag{13.10}$$

For example, we could satisfy these conditions by placing our sample in the resonator near $x = L_x/2; y = 0$. This corresponds to putting the sample at the location of the maximum of the magnetic field, which is polarized along the x axis (see Fig. 38). Next, we assume that the magnetic field B_{ex} in (13.6) includes a homogeneous component, $\vec{B}_0$, directed along the z axis, as well as a time-periodic component, directed along the x axis and given by

$$b(t) = b_0 \cos(\Omega t). \tag{13.11}$$

Thus, b_0 and Ω denote the amplitude and frequency of the time-periodic component of the external magnetic field. We also assume that the following frequency conditions are satisfied

$$\gamma < |\omega - \Omega| \ll \omega, \Omega. \tag{13.12}$$

We shall discuss experimental parameters corresponding to this condition in the Conclusion.

Now, rewriting (13.5) and (13.6) with only one mode (13.8) yields the following Hamiltonian for our system

$$\mathbf{H} = \hbar\omega \mathbf{a}^+\mathbf{a} + g\mu_B B_0 \sum_{j=1}^{N} \mathbf{S}_j^z + \frac{g\mu_B c k_y}{\omega}\sqrt{\frac{2\pi\hbar\omega}{V}} \sum_{j=1}^{n} (\mathbf{S}_j^+ + \mathbf{S}_j^-)(\mathbf{a}^+ + \mathbf{a}) \tag{13.13}$$

$$+\frac{1}{2} g\mu_B b_0 \cos(\Omega t) \sum_{j=1}^{N} (\mathbf{S}_j^+ + \mathbf{S}_j^-).$$

Below, we show that the strength of interaction between the spins and the resonator mode is determined by the real coupling constant

$$\Lambda_0 = \sqrt{\frac{2\pi c^2 k_y^2 \mu_B^2 g^2 N}{\hbar\omega^3 V}}. \tag{13.14}$$

The quantum effects in which we are interested depend significantly on the number of spins, N. Thus, it is convenient to introduce experimentally controllable parameters that would allow us to vary the number of spins, N, at a fixed value of the coupling, Λ_0. For this we choose the linear dimensions of the sample to satisfy

$$l_{x,y} \ll L_{x,y}, \quad l_z = L_z. \tag{13.15}$$

In this case, the field $B_x(x, y)$ in (13.8) may be considered as homogeneous over the size of the sample. The number of spins in the sample is: $N = \rho_0 l_x l_y l_z$ (where ρ_0 is the density of spins); so, the constant Λ_0 does not depend on L_z

$$\Lambda_0 = \sqrt{\frac{2\pi c^2 k_y^2 \mu_B^2 g^2 l_x l_y \rho_0}{\hbar\omega^3 L_x L_y}}. \tag{13.16}$$

At the same time, N is proportional to L_z, so N may be varied at constant Λ_0. Estimates of these parameters will be discussed in the Conclusion.

13.3 Slow Dynamics and Its Time-Independent Hamiltonian

In this section we simplify Hamiltonian (13.13). For this, we use the interaction representation with Hamiltonian

$$\mathbf{H}_0 = \hbar\omega\mathbf{a}^+\mathbf{a} + g\mu_B B_0 \sum_{j=1}^{N} \mathbf{S}_j^z, \qquad (13.17)$$

and introduce "collective operators" (11.8). It is easy to see that in the interaction representation use of (13.13) implies the following Hamiltonian, describing only slow dynamics of the system (dropping nonresonant terms)

$$\tilde{\mathbf{H}}_{int}(\tau) = N[(\mathbf{A}\mathbf{S}^+ + \mathbf{A}^+\mathbf{S}^-) + \lambda(e^{i\bar{\Delta}\tau}\mathbf{S}^- + e^{-i\bar{\Delta}\tau}\mathbf{S}^+)], \qquad (13.18)$$

$$i\frac{\partial\tilde{\Psi}_i(\tau)}{\partial\tau} = \tilde{\mathbf{H}}_{int}(\tau)\tilde{\Psi}_i(\tau).$$

In (13.18) a slow dimensionless time, $\tau = \omega_c t$ and a dimensionless constant of interaction of paramagnetic atoms with the external field have been introduced, where

$$\lambda = \frac{g\mu_B b_0}{4\hbar\omega\Lambda_0}, \quad \omega_c = \omega\Lambda_0, \quad \bar{\Delta} = (\Omega - \omega)/\omega_c. \qquad (13.19)$$

The wave function $\tilde{\Psi}_i(\tau)$ describes the slow dynamics of our system in the interaction representation. In deriving the Hamiltonian (13.18) we assume that $\Lambda_0 \ll 1$, so the dynamics of the system may be separated into slow and fast motion. When the condition $\Lambda_0 \ll 1$ is satisfied, one may neglect the fast terms in the Hamiltonian ($\sim \exp(\pm 2i\omega t)$) and use the rotating wave approximation (RWA). So, in this case we have: $\omega_c/\omega \sim \Lambda_0 \ll 1$. The slow frequency ω_c introduced in (13.19) plays here the role of a "cooperative frequency", that characterizes the time-scale for energy exchange between paramagnetic atoms and the self-consistent field in the resonator.

The Hamiltonian (13.18) coincides with the Hamiltonian (12.13) considered in detail in the previous section. Thus, all results derived in sections 12.3-12.7 can be immediately applied to our magnetic system.

We briefly describe here the main results. In the semiclassical limit (11.34) a qualitative condition for developed chaos to occur in the system may be expressed as (12.47), where in the case considered here $\omega_c = \omega\Lambda_0$ is the cooperative frequency; $\omega_R = g\mu_B b_0/2\hbar$ is the Rabi frequency and $\Delta = \Omega - \omega$ is the detuning. Numerical results for the semiclassical dynamics are shown in Figs. 30-33. In the case of developed semiclassical chaos the dynamics is locally unstable in most regions of phase space. Thus, one could expect considerable influence of quantum corrections on this semiclassical dynamics. As was already shown in sections 12.3-12.7, the time-scale $\tau_\hbar$ of violation of the semiclassical approximation is of the order $\tau_\hbar \sim \ln N$ and the quantum correlation functions grow exponentially during the times: $0 < \tau < \tau_\hbar$. When the semiclassical dynamics is regular, the quantum correlation functions grow significantly slower, as an algebraic power law.

13.4 Conclusion

Materials with strong concentrations of free radicals may be suitable for experimental observation of the quantum effects we discuss here, arising in the classical/quantum transition from nondissipative semiclassical dynamics [106-108]. In the solid state, these materials have rather thin PMR lines, and the value of the g-factor is close to 2. One of the best known of such materials has chemical composition: α,α-diphenyl-B-picryl hydrazyl (DPPH). This material is used as a standard for PMR and has a half-width at half-height $\Delta B/2 \sim 1Oe$. The value ΔB depends weakly on temperature, and on the frequency at which the measurement is made. For example, if we use a solvent from which a sample of DPPH is crystallized, then we have

$$\Delta B/2 = 1.45Oe, \quad for \quad f = 300MHz, \quad T = 295^o K, \tag{13.20}$$

$$\Delta B/2 = 1.3Oe, \quad for \quad f = 9.4GHz, \quad T = 295^o K,$$

$$\Delta B/2 = 1.3Oe, \quad for \quad f = 300MHz, \quad T = 90^o K.$$

The average distance between unpaired electrons in DPPH $\sim 10^{-7} cm$ which corresponds to the density: $\rho_0 = 10^{21} cm^{-3}$. The thin width and large density guarantee a high level of PMR signal, allowing measurements with as little as $\sim 10^{-9} gm$ of DPPH. Note, that in experiments even thinner PMR lines have been observed. For example, DPPH in a solution of carbon bisulphide has $\Delta B/2 = 0.65 Oe$; and Picryl-n-amino-car-bazyl [107] (whose structure is slightly different from DPPH) has $\Delta B/2 = 0.25 Oe$ which corresponds to $\Delta f/2 = 70 KHz$.

Now we give numerical estimates for the interaction constants Λ_0 and λ corresponding to a typical PMR frequency $f = 1 GHz$ and $\rho_0 = 10^{21} cm^{-3}$. Assume $k_y^2 \gg k_x^2$, then $\omega^2 \approx c^2 k_y^2$ and $L_y \approx 15 cm$. Let's put $L_x \approx 40 cm$, then substituting these values into (13.16) and assuming $l_y = 3cm, l_x = 6cm$ gives

$$\Lambda_0 \approx 0.1. \tag{13.21}$$

Varying L_z, for example, from $L_z = 0.4cm$ to $L_z = 40cm$ would vary N from $7 \cdot 10^{21} cm^{-3}$ up to $7 \cdot 10^{23} cm^{-3}$. The value Λ_0 remains constant over these variations. We estimate now the dimensionless value of λ for the parameters chosen above. From (13.19) we have

$$\lambda \approx 10^{-2} b_0(Oe). \tag{13.22}$$

So, as b_0 varies in the region: $b_0 = 10^{-3}$ to 10^2 Oe, the value λ varies in the region: $\lambda = 10^{-5}$ to 1. Finally, the cooperative frequency in this case is given by: $\omega_c = 2\pi f \Lambda_0 \approx 0.63 GHz$. With these parameter values, the conditions for developed semiclassical chaos are: $b_0 \approx 10^2 Oe$; $\Delta \approx \omega_c$.

We envision the following experimental setup. In the resonator an electromagnetic wave is injected with nonresonant frequency Ω. After a time-interval $\tau > Q/\pi f$ in the resonator, oscillations with frequency Ω will be established. The frequency Ω differs from the resonant frequency ω of spins, so the influence of the external field on the spin system is small. At the same instant $t_0 > Q/\pi f$ the electron magnetization is suddenly inclined from z axis. After this, for times

$$\Delta t \sim min\{Q/\pi f, (\pi \Delta f)^{-1}\} \tag{13.23}$$

a dynamical process will occur governed by the Hamiltonian (13.18). We have also assumed that temperature effects are small ($\hbar\omega > T$). This unfortunately leads to rather strict limitations on the temperature: at $f = 10^9 GHz$ the temperature should be $T < 10^{-1o}K$.

So far, the choice of parameters has been restricted to keep the constant Λ_0 in (13.16) sufficiently large ($\Lambda_0 \approx 0.1$). Relaxing this restriction allows the possibility of increasing the resonant frequency ω. So the dimension of the resonator and of the sample could be reduced and the allowed temperature could be increased. The decrease of Λ_0 in this case would not necessarily preclude observation of the quantum effects that violate the semiclassical approximation, but it would weaken the corresponding semiclassical chaos.

14 Weak Quantum Chaos in a System of N Atoms in a Resonant Cavity Interacting with an External Resonant Field

In this section we return to the problem which we started to consider in section 12. However, here we shall study the role of quantum effects in the region of parameters for weak chaos in the semiclassical limit (11.34) (weak interaction of atoms with the external field). In quantum Hamiltonian (12.13) this corresponds to the condition

$$\lambda \ll 1, \tag{14.1}$$

where λ is defined in (12.11), and is a dimensionless constant of interaction of atoms with the external resonant field. We recall that when $\lambda = 0$, the Hamiltonian (12.13) reduces to the completely integrable quantum Dicke model (QDM) [24]. The problem of quantization of weak chaos in the model (12.13) is of particular interest, as weak chaos represents an intermediate stage between integrable (unperturbed) motion, when $\lambda = 0$, and developed chaos ($\lambda \sim 1$), considered in section 12. Weak chaos in the semiclassical limit occurs in the system (12.13) in the vicinity of unperturbed separatrix (see

Figs. 30b,c), and was considered analytically in [103,104]. We also note, that the experimental realization of the conditions of weak quantum chaos in the system with the Hamiltonian (12.13) looks rather promising.

The main results of our consideration are similar to the results of sections 9.7b and 12. We show that in the region of parameters of weak chaos in the semiclassical approximation, under the conditions $N \gg 1$, and for a rather narrow initial packet: $\sigma_{s^z} < \sqrt{N}$, quantum effects lead to violation of the semiclassical description on the time-scale $\tau_\hbar \sim \ln N$. Moreover, QCFs of the type (11.35) grow exponentially in time on the time-interval $0 < \tau < \tau_\hbar$, and this effect of enhancement of QCFs may be experimentally measurable.

In fact, the problem discussed in this section is close to the same problem in the quantum Dicke model (QDM), discussed in [24,105] (see also section 14.4). Although QDM is a completely integrable system, more then twenty years ago results for this model were obtained [105] that forshadow how the time-scale $\tau_\hbar \sim \ln N$ for violation of the semiclassical approximation arises in a system governed by the Hamiltonian

$$\mathbf{H}_D = N\left[\mathbf{A}\mathbf{S}^+ + \mathbf{A}^+\mathbf{S}^-\right], \tag{14.2}$$

which follows from (12.13) when $\lambda = 0$, and corresponds to QDM.

The authors of [105] study the quantum dynamics and corresponding classical motion in QDM, starting from an initial "super radient" state, and from an initially completely inverted state (CIS). For CIS, they found that a difference between quantum slow dynamics and its corresponding classical limit first appears on a time-scale, $\tau_\hbar \sim \ln N$.

We redard our results as an extension of the results derived in [105] for the case of chaotic semiclassical dynamics. Namely, our statement is the following: if in the semiclassical approximation there exists chaos (either weak, or developed, as considered in section 12), then $\tau_\hbar \approx \ln N$ will be the characteristic time after which significant differences may arise between classical and quantum dynamics. Roughly speaking, one can say that the logarithm appears in $\tau_\hbar$ because of local exponential instability of classical trajectories in phase space, and N appears because N is the only "quantum" parameter appearing in equations (12.25)-(12.28).

14.1 The Nonsingular Representation in the Vicinity of Separatrix

Recall equations (12.25)-(12.28), which describe the dynamics of an arbitrary quantum expectation value $f(\tau)$ for a system modelled by Hamiltonian (12.13) or (12.19).

In this section we are interested mainly in the dynamics of the system initially populated in the vicinity of separatrix, where the atomic (spin) sub-system can be completely inverted (see (11.33))

$$|\xi| \to \infty, \quad S^z \equiv < \mathbf{S}^z(0) > \to \frac{J}{N}, \quad S^{\pm} \equiv < \mathbf{S}^{\pm}(0) > \to 0.$$

As one can see from (12.28), in this case the quantum operator $\hat{K}_q$ in (12.28) includes singular terms (we omit the superscript 1 in the operators). So, it is convenient to transform to new axes of quantization, where this formal singularity vanishes. The new axes are related to the old ones by a rotation about the x axis, by the angle π. The spin operators transform as vectors under this rotation, namely

$$\mathbf{S}^{z\prime} = -\mathbf{S}^z, \quad \mathbf{S}^{x\prime} = \mathbf{S}^x, \quad \mathbf{S}^{y\prime} = -\mathbf{S}^y, \tag{14.3}$$

$$\mathbf{J}^{z\prime} = -\mathbf{J}^z, \quad \mathbf{J}^{x\prime} = \mathbf{J}^x, \quad \mathbf{J}^{y\prime} = -\mathbf{J}^y.$$

The transformation (14.3) leads to the transformation of variables in the Hamiltonian (12.19)

$$\boldsymbol{\eta}^z \to -\boldsymbol{\eta}^z, \quad \boldsymbol{\eta}^+ \to \boldsymbol{\eta}^-, \quad \boldsymbol{\eta}^- \to \boldsymbol{\eta}^+. \tag{14.4}$$

In the new axes the Hamiltonian (12.19) takes the form

$$\mathbf{H} = N\left\{\mathbf{c}\boldsymbol{\eta}^- + \mathbf{c}^+\boldsymbol{\eta}^+ + \lambda\boldsymbol{\eta}^+ + \lambda\boldsymbol{\eta}^- - \bar{\Delta}\mathbf{c}^+\mathbf{c} + \bar{\Delta}\boldsymbol{\eta}^z\right\}. \tag{14.5}$$

From (14.15) we derive the Heisenberg equations of motion,

$$\dot{\mathbf{c}} = i\bar{\Delta}\mathbf{c} - i\boldsymbol{\eta}^+, \tag{14.6}$$

$$\dot{\boldsymbol{\eta}}^{+} = i\bar{\Delta}\boldsymbol{\eta}^{+} - 2i\boldsymbol{\eta}^{z}\mathbf{c} - 2i\lambda\boldsymbol{\eta}^{z},$$
$$\dot{\boldsymbol{\eta}}^{z} = i\mathbf{c}\boldsymbol{\eta}^{-} - i\mathbf{c}^{+}\boldsymbol{\eta}^{+} + i\lambda\boldsymbol{\eta}^{-} - i\lambda\boldsymbol{\eta}^{+}.$$

We introduce new spin coherent states

$$|\mu >= (1+|\mu|^2)^{-J}\exp[\mu(\mathbf{J}^{+})']||J,-J>>, \quad (\mathbf{J}^{+})' = (\mathbf{J}^{x})' + i(\mathbf{J}^{y})', \tag{14.7}$$

where $||J,m>>$ is the eigenstate of the operator $(\mathbf{J}^z)'$

$$(\mathbf{J}^z)'||J,m>>= m||J,m>> . \tag{14.8}$$

As seen from (14.3), we have

$$||J,m>>= |J,-m> .$$

Then, (14.3) and (14.7) imply

$$|\mu >= (1+|\mu|^2)^{-J}\exp(\mu\mathbf{J}^{-})|J,J> . \tag{14.9}$$

In the new coordinate system we have the analog of equations (12.25)-(12.28)

$$\frac{\partial f(\tau)}{\partial \tau} = \hat{K}f(\tau). \tag{14.10}$$

The operator $\hat{K}$ has the form

$$\hat{K} = \hat{K}_{cl} + \frac{1}{N}\hat{K}_q, \tag{14.11}$$

where

$$\hat{K}_{cl} = i[-\bar{\Delta}z^*\frac{\partial}{\partial z^*} + \bar{\Delta}z\frac{\partial}{\partial z} + \bar{\Delta}\mu^*\frac{\partial}{\partial \mu^*} - \bar{\Delta}\mu\frac{\partial}{\partial \mu} \tag{14.12}$$

$$+\frac{2J}{N}\frac{1}{(1+|\mu|^2)}(\mu\frac{\partial}{\partial z^*} - \mu^*\frac{\partial}{\partial z}) + (z\mu^2\frac{\partial}{\partial \mu} - z^*\mu^{*2}\frac{\partial}{\partial \mu^*}) + (z\frac{\partial}{\partial \mu^*} - z^*\frac{\partial}{\partial \mu})$$
$$+\lambda(\frac{\partial}{\partial \mu^*} - \frac{\partial}{\partial \mu} + \mu^2\frac{\partial}{\partial \mu} - \mu^{*2}\frac{\partial}{\partial \mu^*})],$$

and

$$\hat{K}_q = i\left(\frac{\partial^2}{\partial z^*\partial \mu^*} - \frac{\partial^2}{\partial z\partial \mu}\right). \tag{14.13}$$

One can see that the values ξ in (12.27), (12.28) and μ in (14.12), (14.13) are related

$$\xi = 1/\mu. \tag{14.14}$$

As in (12.26), (12.27), the operator $\hat{K}_{cl}$ in (14.12) describes the semiclassical motion, and the operator $\hat{K}_q$ (14.13) describes quantum effects. As one can see, the operator $\hat{K}_q$ does not include singular terms.

14.2 Equations for Quantum Expectation Values

Using QCFs of the type (11.35) for various operators **f** and **g**, and the operator equations (14.6) for different operators **f** and **g**, leads to the following set of *c*-number equations for expectation values

$$\dot{c} = i\bar{\Delta}c - i\eta^+, \tag{14.15}$$

$$\dot{\eta}^+ = i\Delta\eta^+ - 2i\lambda\eta^z - 2i\eta^z c - \frac{2i}{N}P_{\eta^z,c},$$

$$\dot{\eta}^z = ic\eta^- - ic^+\eta^+ + i\lambda\eta^- - i\lambda\eta^+ + \frac{i}{N}(P^*_{c^+,\eta^+} - C.C.).$$

The expectation values for the integrals of motion in the new variables are given by

$$E_q \equiv < \mathbf{H} >= c\eta^- + c^+\eta^+ + \lambda\eta^+ + \lambda\eta^- - \bar{\Delta}c^+c + \bar{\Delta}\eta^z \tag{14.16}$$

$$+\frac{1}{N}(P_{c^+,\eta^+} + C.C.) - \frac{\bar{\Delta}}{N}P_{c^+,c},$$

$$W \equiv < \mathbf{W} >= -\eta^z + c^+c + \frac{1}{N}P_{c^+,c},$$

$$< \mathbf{S}^2 >= (\eta^z)^2 + \eta^+\eta^- + \frac{1}{N}P_{\eta^z,\eta^z} + \frac{1}{2N}(P_{\eta^+,\eta^-} + C.C.).$$

In (14.16) W is an integral of motion only if $\lambda = 0$.

Before considering the quantum effects when the system is initially populated in the vicinity of the separatrix, we consider the corresponding semiclassical dynamics, which is described by the operator $\hat{K}_{cl}$ (14.12).

14.3 Semiclassical Consideration

The semiclassical dynamics for this system has already been sketched in section 12.5, and in the case of weak chaos it is treated in detail in [103,104]. Consequently, we present here only the corresponding equations in the new variables. Instead of (12.45), we have now the following equations for characteristics

$$\frac{dz}{d\tau} = i\bar{\Delta}z - i\frac{g\mu^*}{(1+|\mu|^2)}, \quad g = \frac{2J}{N}, \tag{14.17}$$
$$\frac{d\mu}{d\tau} = -i\bar{\Delta}\mu - iz^* + iz\mu^2 - i\lambda + i\lambda\mu^2,$$

where g takes values: g=0,1/N,...,1. From (14.17) we have the semiclassical solutions for atomic (spin) variables

$$\eta^+(\tau) = \frac{g\mu^*(\tau)}{(1+|\mu(\tau)|^2)}, \quad \eta^z(\tau) = -\frac{g(1-|\mu(\tau)|^2)}{2(1+|\mu(\tau)|^2)}. \tag{14.18}$$

Equations (14.17) may also be written in Hamiltonian form, as

$$\dot{z} = -i\frac{\partial H_{cl}}{\partial z^*}, \quad \dot{\zeta} = -i\frac{\partial H_{cl}}{\partial \zeta^*}, \quad \left(\zeta = \frac{\sqrt{g}\mu}{\sqrt{1+|\mu|^2}}\right), \tag{14.19}$$

with the following semiclassical Hamiltonian

$$H_{cl} = \bar{\Delta}(|\zeta|^2 - |z|^2) + \sqrt{g - |\zeta|^2}[z\zeta + z^*\zeta^* + \lambda(\zeta + \zeta^*)]. \tag{14.20}$$

We present the results of numerical calculations on weak chaos in the system with the Hamiltonian (14.5) in section 14.5. In the next section we present results connected with the $\ln N$ time-scale, which appear already in the integrable limit (14.1).

14.4 Dynamics of Expectation Values and Quantum Correlation Functions in the Integrable Limit

In this section we show how the time-scale $\tau_\hbar \sim \ln N$ arises in dynamics of expectation values and QCFs in the quantum Dicke model with the Hamiltonian (14.2). Using the commutation relations (11.8), we derive from Hamiltonian (14.2) the following operator equations

$$\dot{\mathbf{A}} = -i\mathbf{S}^-, \tag{14.21}$$
$$\dot{\mathbf{S}}^- = 2i\mathbf{A}\mathbf{S}^z,$$
$$\dot{\mathbf{S}}^z = i(\mathbf{A}^+\mathbf{S}^- - \mathbf{A}\mathbf{S}^+).$$

First, consider the system of equations (14.21) in the semiclassical limit (11.34). In this case all operators in (14.21) can be replaced by c-numbers. Using the substitution $\alpha = iA$, we have from (14.21) in the semiclassical approximation

$$\dot{\alpha} = S^-, \tag{14.22}$$
$$\dot{S}^- = 2\alpha S^z,$$
$$\dot{S}^z = -\alpha S^+ - \alpha^* S^-.$$

As one can see, the substitution

$$R^+(\tau) = e^{-i\Psi} S^+(\tau), \quad \beta(\tau) = e^{i\Psi}\alpha(\tau), \tag{14.23}$$

does not change the system of equations (14.22), namely

$$\dot{\beta} = R^-, \quad \dot{R}^- = 2\beta S^z, \quad \dot{S}^z = -\beta R^+ - \beta^* R^-. \tag{14.24}$$

The initial conditions for the system of equations (14.24) are the following

$$S^z(0) = S^z, \quad R^-(0) = e^{-i\Psi} S^-(0), \quad R^+(0) = e^{i\Psi} S^+(0), \tag{14.25}$$
$$\beta(0) = e^{i\Psi}\alpha(0).$$

It is clear that an arbitrary solution of the system (14.24), (14.25) will be a solution of the system (14.22). We introduce new variables,

$$R^+ = R_x + iR_y, \quad \beta = \beta_x - i\beta_y. \tag{14.26}$$

In these variables the system of equations (14.24) takes the form

$$\dot{\beta}_x = R_x, \quad \dot{\beta}_y = R_y, \tag{14.27}$$

$$\dot{R}_x = 2\beta_x S^z, \quad \dot{R}_y = 2\beta_y S^z, \quad \dot{S}^z = -2\beta_x R_x - 2\beta_y R_y.$$

From (14.27) it follows that if $R_y(0) = \beta_y(0) = 0$, then $R_y(\tau) = \beta_y(\tau) = 0$, and we have in this case the following system of equations

$$\dot{\beta}_x = R_x, \quad \dot{R}_x = 2\beta_x S^z, \quad \dot{S}^z = -2\beta_x R_x, \tag{14.28}$$

which under the condition $\beta(0) = 0$ describes the general solution of the initial system (14.22). Using the substitution

$$R_x = \sin\varphi, \quad S^z = \cos\varphi, \tag{14.29}$$

we derive from (14.28)

$$\dot{\varphi} = 2\beta_x, \quad \ddot{\varphi} = 2\sin\varphi. \tag{14.30}$$

The equation for the phase φ in (14.30) coincides with the well known equation for the mathematical pendulem. From (14.30) it follows that if at $\tau = 0$ the radiation field is in vacuum state ($\beta_x(0) = 0$), and atomic (spin) sub-system is completely inverted ($\varphi(0) = \dot{\varphi}(0) = 0$), then, the system is at a hyperbolic equilibrium point, and the corresponding semiclassical dynamics will be unstable.

We now consider the quantum dynamics for the Hamiltonian (14.2), when at $\tau = 0$ the radiation field is in vacuum state $|n = 0>$, and the atomic sub-system is completely inverted [105]. We write the Schrödinger equation in the form

$$i\frac{\partial|\Psi>}{\partial\tau} = \mathbf{H}_D|\Psi> = \frac{1}{\sqrt{N}}[\mathbf{a}\mathbf{J}^+ + \mathbf{a}^+\mathbf{J}]|\Psi>, \tag{14.31}$$

where $\mathbf{J} = N\mathbf{S}$, and $\mathbf{A} = \mathbf{a}/\sqrt{N}$. Suppose at $\tau = 0$ the atomic sub-system is completely inverted with cooperative number J. The cooperative number J can take the values: $J = 0, 1/2, ..., N/2$. However, since we are interested in comparing quantum and classical dynamics, we shall consider $J \sim N$. The dynamics of this initial state is described by the following wave function [105]

$$|\Psi(\tau)> = \sum_{n=0}^{2J} c(n,\tau)|n>|J, J-n>, \tag{14.32}$$

where $|n>$ is a photon state, and $|J, J-n>$ is an atomic (spin) state with the projection $J-n$ on the axis z.

From (14.32) we approximate the probability amplitude for finding n photons at the time τ, as

$$c(n,\tau) \equiv < n|\Psi(\tau)> \approx (-i)^n \tanh^n \tau \cdot \sec \tau. \tag{14.33}$$

The approximation (14.33) can be used only for finite times, when the following condition is satisfied

$$\bar{n}(\tau) \ll 2J, \quad or \quad \left(1 - \frac{< \mathbf{J}^z >}{J}\right) \ll 1, \tag{14.34}$$

where $\bar{n}(\tau)$ is the average number of photons. In this approximation, the probability of finding n photons is given by Bose statistics [105]

$$p(n,\tau) = |c(n,\tau)|^2 = \frac{\bar{n}^n}{(1+\bar{n})^{n+1}}, \tag{14.35}$$

with average number of photons $\bar{n}$, and standard deviation σ_n

$$\bar{n} \equiv \bar{n}(\tau) = \sinh^2 \tau \sim e^{2\tau}, \tag{14.36}$$

$$\sigma_n = \sqrt{\bar{n^2} - \bar{n}^2} = \sqrt{\bar{n}(\bar{n}+1)} \approx \bar{n} = \sinh^2 \tau \sim e^{2\tau}.$$

Consider the quantum correlation function (QCF)

$$\mathcal{P}_{J^z,J^z} = \frac{1}{J^2}[< (\mathbf{J}^z)^2 > -(< \mathbf{J}^z >)^2] \tag{14.37}$$

$$= \frac{1}{J^2(1+\bar{n})} \sum_{n=0}^{2J} \left[\frac{\bar{n}}{1+\bar{n}}\right]^n (J-n)^2$$

$$- \frac{1}{J^2(1+\bar{n})} \left[\sum_{n=0}^{2J} \left[\frac{\bar{n}}{1+\bar{n}}\right]^n (J-n)\right]^2$$

$$\approx \frac{\bar{n}+\bar{n}^2}{J^2} \sim \frac{e^{4\tau}}{J^2}, \quad (J \gg 1, \quad \sinh^2 \tau/2J \ll 1).$$

Thus, the QCF which characterizes the standard deviation of the expectation value $< \mathbf{J}^z >$ grows exponentially in the time-interval $0 < \tau < \tau_\hbar \sim \ln N$. Also, the following estimate holds

$$1 - \frac{< \mathbf{J}^z >}{J} \approx \frac{\bar{n}}{J} = \frac{\sinh^2 \tau}{J} \sim \frac{e^{2\tau}}{J}. \qquad (14.38)$$

From (14.36)-(14.38) we have the following useful formula for $\mathcal{P}_{J^z,J^z}$

$$\mathcal{P}_{J^z,J^z} \sim \left(1 - \frac{< \mathbf{J}^z >}{J}\right)^2. \qquad (14.39)$$

Now we present the results of numerical calculations for the QDM (14.2).

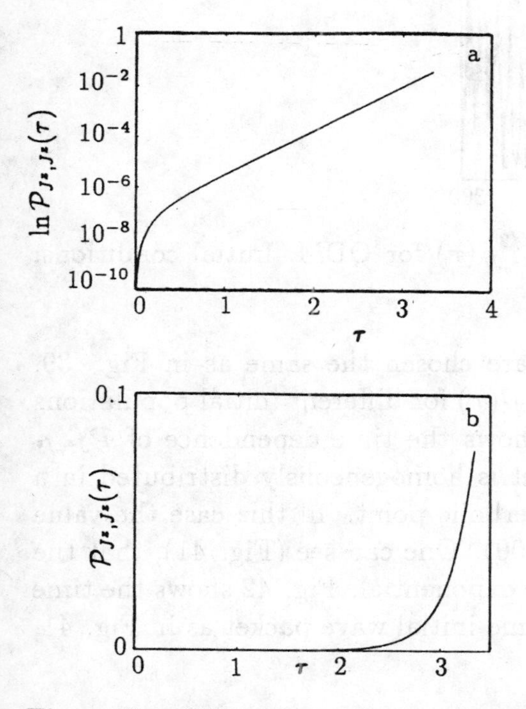

Fig. 39. Time dependence of QCF $\mathcal{P}_{J^z,J^z}(\tau)$ (14.37) for quantum Dicke model (QDM); $\mathcal{P}_{J^z,J^z}(0) = 0$. Initial wave function: $|\Psi(0) = |0> |J, J>$; $J = 10^3$; (a) $\ln \mathcal{P}_{J^z,J^z}(\tau)$; (b) $\mathcal{P}_{J^z,J^z}(\tau)$.

Fig. 39 shows the time behavior of QCF $\mathcal{P}_{J^z,J^z}(\tau)$ (14.37). At $\tau = 0$ the atomic sub-system is completely inverted, and the radiation field is in the vacuum state. As one can see, the time behavior of QCF $\mathcal{P}_{J^z,J^z}(\tau)$ in this case follows an exponential law. This result is rather important, as the QCF $\mathcal{P}_{J^z,J^z}(\tau)$ actually gives the standard deviation for the expectation value $< \mathbf{J}^z >$, and usually has a power-law dependence on time (of the diffusion type). Fig. 40 shows the long time behavior of QCF $\mathcal{P}_{J^z,J^z}(\tau)$.

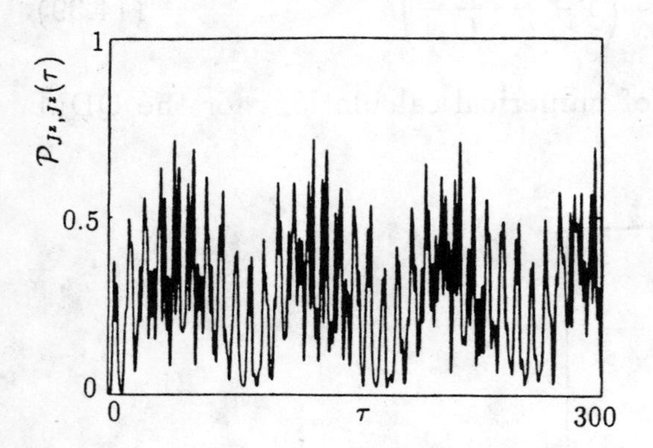

Fig. 40. Dependence QCF $\mathcal{P}^{1/2}_{J^z,J^z}(\tau)$ for QDM. Initial conditions: $|\Psi(0) >= |0 > |J, J >$; $J = 10^2$.

Initial conditions in this case are chosen the same as in Fig. 39. Also we compute the QCF $\mathcal{P}_{J^z,J^z}(\tau)$ for different initial populations of the wave packet. Fig. 41 shows the time dependence of $\mathcal{P}_{J^z,J^z}$ when at $\tau = 0$ the wave packet is homogeneously distributed in a cone in the vicinity of the hyperbolic point. In this case the value of J is chosen as $J = 50(N = 100)$. One can see (Fig. 41), that the dependence $\mathcal{P}_{J^z,J^z}(\tau)$ on time is exponential. Fig. 42 shows the time behavior of $\mathcal{P}_{J^z,J^z}(\tau)$ for the same initial wave packet as in Fig. 41, but for $J = 310(N = 620)$.

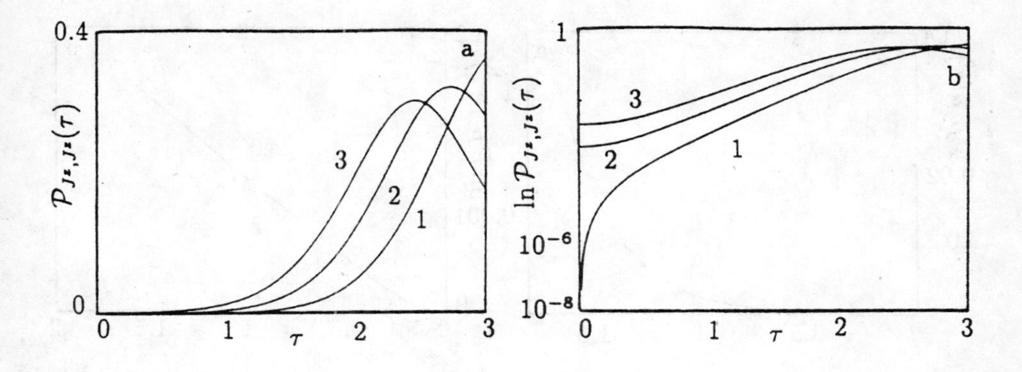

Fig. 41. Dependence $\mathcal{P}_{J^z,J^z}(\tau)$ for QDM. Initial wave function (14.32): $c(n,0) = 1/\sqrt{n_0}; 0 \leq n < n_0 = xJ$, and $c(n \geq n_0, 0) = 0$; (a) $J = 50$; curve 1: $x = 0$; curve 2: $x = 0.5$; curve 3: $x = 1$; (b) $\ln \mathcal{P}_{J^z,J^z}(\tau)$. Parameters the same as in (a).

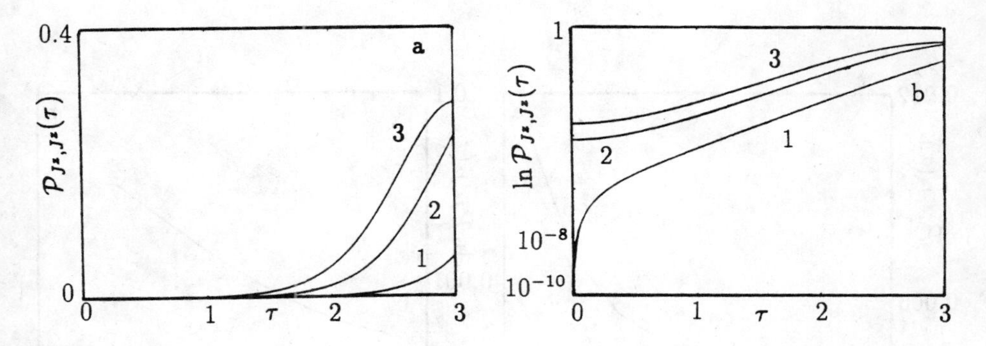

Fig. 42. Dependence $\mathcal{P}_{J^z,J^z}(\tau)$ for QDM. Initial wave function (14.32): $c(n,0) = 1/\sqrt{n_0}; 0 \leq n < n_0 = xJ$, and $c(n \geq n_0, 0) = 0$; (a) $J = 310$; curve 1: $x = 0$; curve 2: $x = 0.5$; curve 3: $x = 1$; (b) $\ln \mathcal{P}_{J^z,J^z}(\tau)$. Parameters the same as in (a).

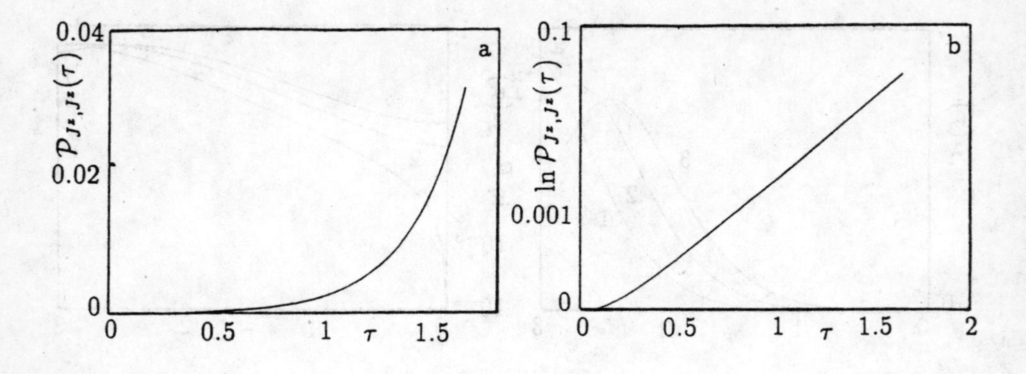

Fig. 43. Dependence $\mathcal{P}_{J^z,J^z}(\tau)$ for QDM. Initial wave function (14.32) is chosen in the form (14.40); (a) $J = 50$; (b) $\ln \mathcal{P}_{J^z,J^z}(\tau)$. Parameters the same as in (a).

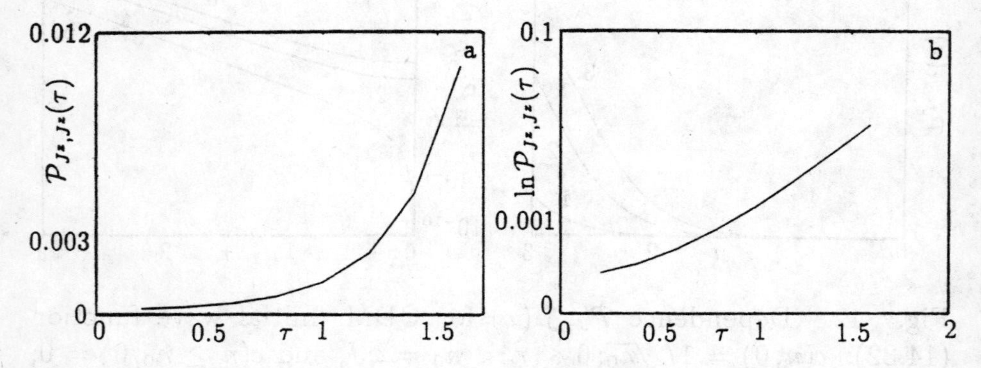

Fig. 44. Dependence $\mathcal{P}_{J^z,J^z}(\tau)$ for QDM. Initial wave function (14.32) is chosen in the form (14.40); (a) $J = 5 \cdot 10^3$; (b) $\ln \mathcal{P}_{J^z,J^z}(\tau)$. Parameters the same as in (a).

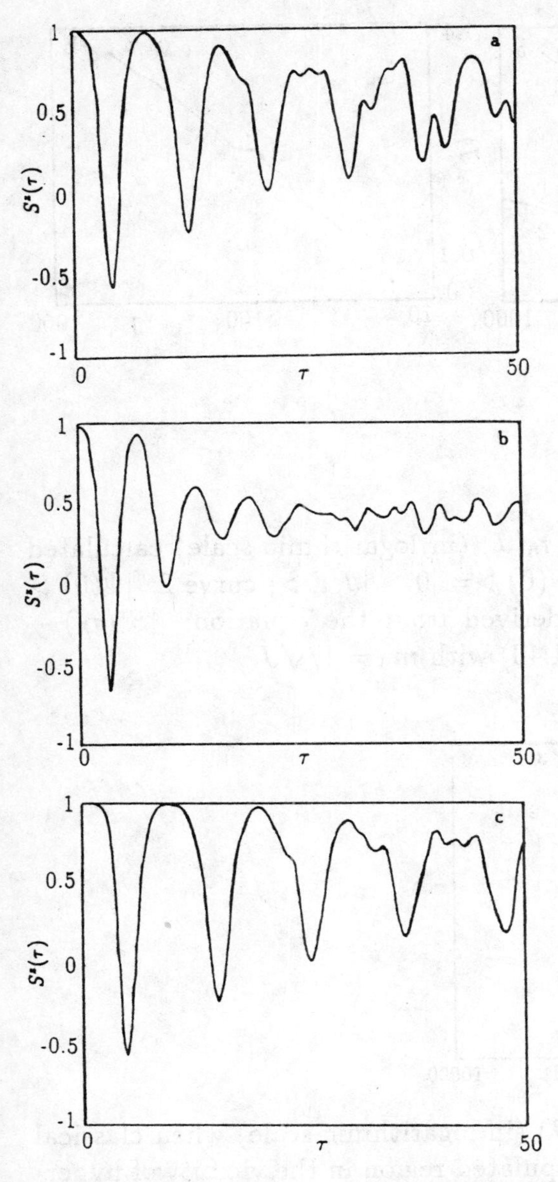

Fig. 45. Dependence $S^z(\tau)$ for QDM; (a) $|\Psi(0)>=|0>|J,J>$; $J=100$; (b) $|\Psi(0)>$ (14.43) with $n_0=50$; $J=100$; (c) $|\Psi(0)>=|0>|J,J>$; $J=500$.

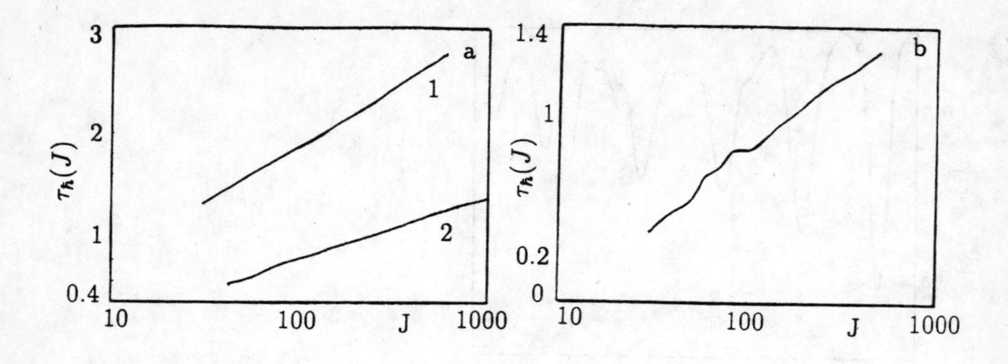

Fig. 46. Time dependence $\tau_\hbar(J)$ (in logarithmic scale) calculated from (14.41); (a) curve 1: $|\Psi(0)>=|0>|J,J>$; curve 2: $|\Psi(0)>$ (14.43). Time-scale $\tau_\hbar$ is derived from the equation: $|S^z(\tau_\hbar)-S^z(0)|=0.1$; (b) $|\Psi(0)>$ (14.40) with $m_0=1/\sqrt{J}$.

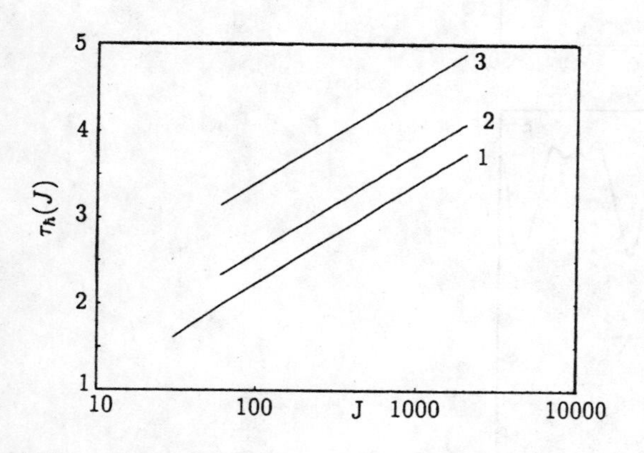

Fig. 47. The time-scale $\tau_{cl}(J)$ (in logarithmic scale) when classical packet escapes an initially populated region in the vicinity of hyperbolic point. Initial packet is homogeneously distributed in the cone with the angle φ (14.44). Curve 1: $x=1$; curve 2: $x=0.5$; curve 3: $x=0.1$.

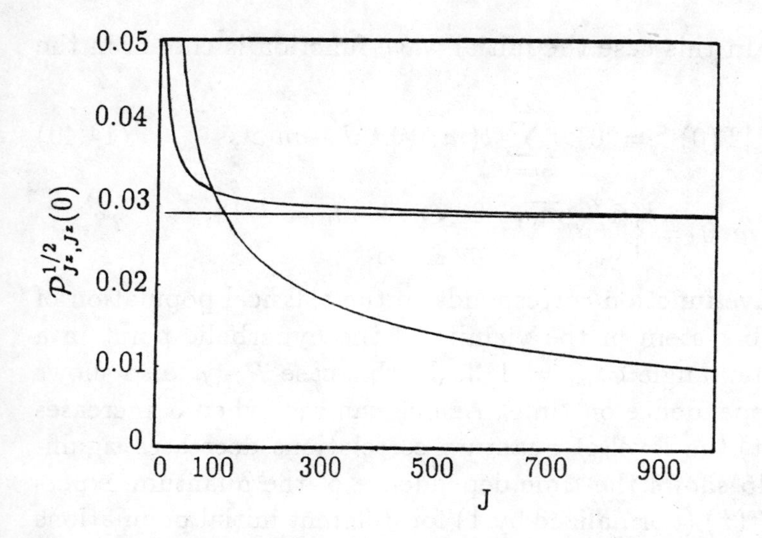

Fig. 48. Dependence of the standard deviation of the expectation value J^z: $\mathcal{P}^{1/2}_{J^z,J^z}(0)$ on J for different *initial* wave packets. Curve 1: classical wave packet with constant initial standard deviation; curve 2: standard deviation for the corresponding initial coherent state; curve 3: standard deviation of the initial quantum wave packet.

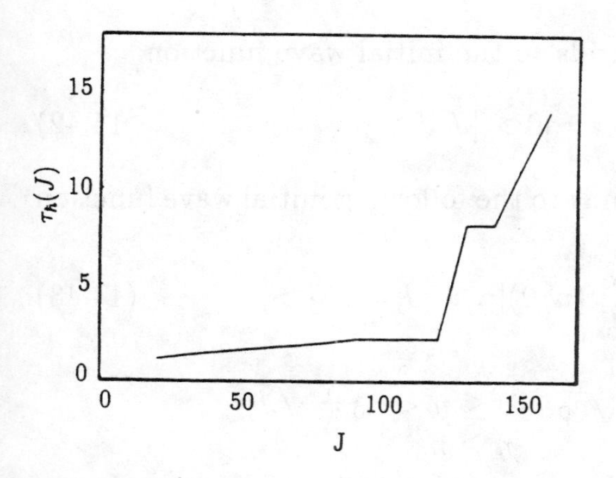

Fig. 49. Crossover in the dependence $\tau_\hbar(J)$ when the width of the initial wave packet goes from narrow to the wide.

Now we present results on the dynamics of QCF $\mathcal{P}_{J^z,J^z}(\tau)$ for initial conditions implying different cooperative numbers J (see Fig. 43

and Fig. 44). In this case the initial wave function is chosen in the form

$$|\Psi(0)>=|0>\sum_{m=0}^{2J} b(m,0)|J,J-m>, \tag{14.40}$$

$$b(m,0)=\begin{cases}1/\sqrt{m_0}, & 0\le m<m_0=J/18,\\ 0; & m\ge m_0.\end{cases}$$

This initial wave function corresponds to the classical population of the atomic sub-system in the vicinity of the hyperbolic point in a cone with vertex angle $\varphi_{max}=1/3$. In this case $\mathcal{P}_{J^z,J^z}$ also shows exponential dependence on time. As one can see, when J increases (classical limit) the level of quantum correlations decreases significantly. Fig. 45 shows the time dependence of the quantum expectation value $S^z(\tau)$ (normalized by 1) for different initial populations of the wave packet. Using this function $S^z(\tau)$ we calculate the time-scale $\tau_\hbar$ (see Fig. 46), defined as the characteristic time at which quantum dynamics differs from its classical counterpart. Here we derived $\tau_\hbar$ from the following equation

$$|S^z(\tau_\hbar)-1|=0.1. \tag{14.41}$$

Curve 1 in Fig. 46a corresponds to the initial wave function

$$|\Psi(0)>=|0>|J,J>. \tag{14.42}$$

Curve 2 in Fig. 46a corresponds to the following initial wave function

$$|\Psi(0)>=\sum_{n=0}^{2J} c(n,0)|n>|J,J-n>, \tag{14.43}$$

$$c(n,0)=\begin{cases}1/\sqrt{n_0}, & 0\le n<n_0=\sqrt{J},\\ 0; & n\ge n_0.\end{cases}$$

Fig. 46b corresponds to the initial population of the wave packet in the form (14.40) with $m_0=\sqrt{J}$. As seen in Fig. 46, the dependence of $\tau_\hbar$ on J is close to logarithmic. Fig. 47 shows the time-scale $\tau_{cl}(J)$ at which the classical packet escapes an initially populated region, for different widths of the initial distribution. In this case the classical initial packet is homogeneously distributed in the vicinity of the hyperbolic point in a cone with vertex angle

$$\varphi^2_{max} = \frac{2x}{\sqrt{J}}; \quad (x = 1; 0.5; 3) \tag{14.44}$$

One can see that when the initial classical packet becomes narrow, the characteristic time-scale τ_{cl} considerably increases. This kind of behavior is completely different from the corresponding quantum dynamics (see Fig. 46).

Finally, we present here the results on the *crossover* in the dependence $\tau_{\hbar}(J)$, which appears when considering initial wave packets of different widths (narrow and wide). This numerical experiment is done in the following way. We prepare at $\tau = 0$ several initial wave packets (see Fig. 48) with different standard deviations in the expectation value $< \mathbf{J}^z(0) >$: $\mathcal{P}_{J^z,J^z}(0)$ (see (14.37)). The value of $\mathcal{P}_{J^z,J^z}(0)$ depends on J, and this dependence is shown in Fig. 48. Curve 1 corresponds to an initial classical wave packet, which is chosen with constant $\mathcal{P}^{(cl)}_{J^z,J^z}(0)$ for all values of J. This is, we populate the initial classical packet homogeneously in the vicinity of the hyperbolic point in a cone with vertex angle $\varphi^2_{max}/2 = 0.1$. In this particular case $\mathcal{P}^{(cl)}_{J^z,J^z}(0) = 0.03$ (see Fig. 48, curve 1). The number of trajectories used in this case is: $M = 100$. The curve 2 in Fig.48 shows the dependence of the standard deviation $\mathcal{P}^{(coh)}_{J^z,J^z}(0)$ on J for the corresponding coherent state. In this case we have according to (2.59)

$$\bar{p} = J - < \mathbf{J}^z(0) > = \frac{\varphi^2_{max}}{2} J, \tag{14.45}$$

$$\mathcal{P}^{(coh)}_{J^z,J^z}(0) = \frac{\sqrt{\bar{p}}}{J} = \sqrt{\frac{\varphi^2_{max}}{2J}} = \frac{1}{\sqrt{10J}}.$$

Finally, the quantum initial wave packet was chosen in the form (14.40) with $m_0 = J/10$, as in this case we have

$$\frac{\varphi^2_{max}}{2} = \frac{J - < \mathbf{J}^z_{min}(0) >}{J} = \frac{m_0}{J} = 0.1. \tag{14.46}$$

The dependence of the standard deviation $\mathcal{P}^{(q)}_{J^z,J^z}(0)$ on J is shown in this case in Fig. 48, curve 3.

As one can see from Fig. 48, for such initial populations the quantum wave packet can be considered as narrow for $J < 100$ and wide, for $J > 100$. The second step is the calculation of the time-scale $\tau_\hbar$ for these initial wave packets, for all values of J presented in Fig. 48. The dependence $\tau_\hbar(J)$ is calculated as the solution of the following equation

$$| < \mathbf{J}^z(\tau) > - \bar{J}^z_{cl}(\tau)|_{\tau=\tau_\hbar} = 0.1J. \tag{14.47}$$

The results of these calculations are shown in Fig. 49. As one can see, the dependence $\tau_\hbar(J)$ has a crossover approximately at $J_{cr} \sim 100$, which corresponds to the region $J \sim 100$ in Fig. 48 where the width of the initial quantum wave packet goes from narrow to wide.

The conclusion we make from these numerical calculations is the following. To observe the logarithmic type dependence $\tau_\hbar(J)$ in the completely integrable QDM system (14.2), one needs to choose the initial wave packet to have a rather narrow standard deviation. Also, the crossover in the dependence $\tau_\hbar(J)$ presented in Fig. 49, may be experimentally observable.

14.5 Numerical Calculations of Weak Quantum Chaos

We present here the results of numerical calculations for the case of weak quantum chaos in the system with Hamiltonian (14.5) under the condition (14.1) of weak interaction of atoms with the external coherent field. In this case, in numerical experiments we integrate the equations (14.15) for time dependent expectation values, and the corresponding equations for QCFs presented in (14.15). Initially coherent states are chosen for both the atomic and field sub-systems. Recall that equations (14.15) are written in the inverted coordinate system (see (14.3)), which considerably simplifies the numerical calculations. At the same time all results presented below in this section are given in the initial system of coordinates, the one used in section 12. When we calculate the time-scale $\tau_\hbar$ and QCFs $P(\tau)$, we use the formulas (12.56)-(12.58). The semiclassical motion in the vicinity of separatrix is shown in Figs. 31 for different valuesof $\lambda \ll 1$.

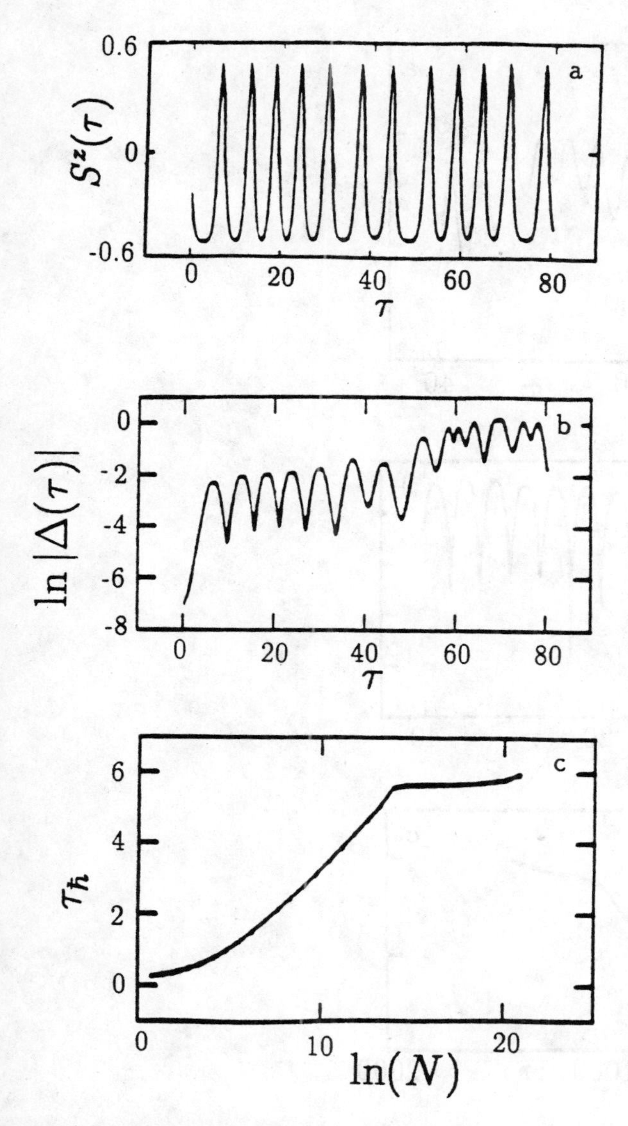

Fig. 50. Weak chaos in the vicinity of separatrix ($g = 1; \bar{\Delta} = 1; \lambda = 0.01$). Initial conditions: $E = 0; W(0) = 0; z(0) = 0.5; \mu(0) \approx 0.58i$; (a) dependence $S^z(\tau)$ in semiclassical limit; (b) dependence $\ln|\Delta(\tau)|$ in semiclassical limit; (c) dependence $\tau_\hbar$ on $\ln N$ derived by "1%-criterion"; $N_{max} = 2^{30} \approx 10^9$.

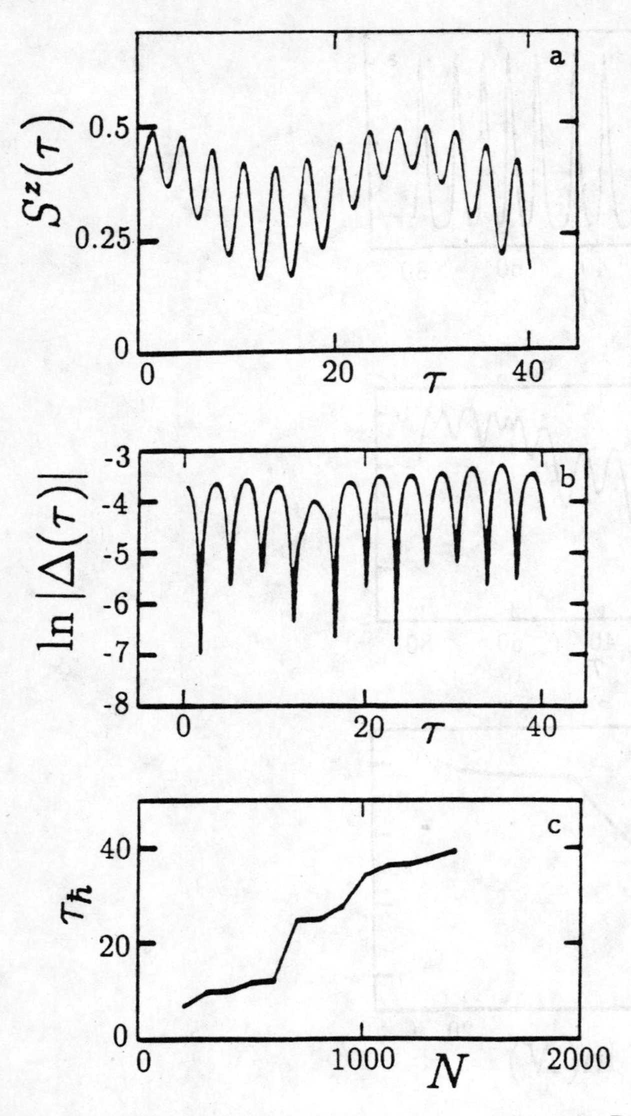

Fig. 51. Dynamics in stable region ($g = 1; \bar{\Delta} = 1; \lambda = 0.01$). Initial conditions: $z(0) = 0; \mu(0) = 3$; (a) dependence $S^z(\tau)$ in semiclassical limit; (b) dependence $\ln|\Delta(\tau)|$ in semiclassical limit; (c) dependence $\tau_\hbar$ on N derived by "1%-criterion".

The semiclassical chaos can be conventially called weak when $\lambda < 0.03$. Figs. 50a,b show the characteristic chaotic semiclassical motion in the vicinity of separatrix. In Fig. 50a the time dependence of semiclassical function $S^z(\tau)$ is shown. In Fig. 50b the time evolution of the distance between this trajectory and one lying near it at $\tau = 0$ is presented. As one can see, the trajectory shown in Fig. 50a is locally unstable. Fig. 50c shows the dependence $\tau_\hbar(N)$. This dependence was derived analogously to the scheme discussed in sections 11.5 and 12.6, by "1% – *criterion*". As one can see from Fig. 50c, the dependence $\tau_\hbar(N)$ in the region $1 \ll N < 10^8$ can be roughly approximated as logarithmic (11.52). For $N > 10^8$ additional numerical calculation are required. Figs. 51a,b shows the results of numerical calculations for the case of stable dynamics for initial conditions far from separatrix. As one can see, the dependence $\tau_\hbar(N)$ (Fig. 51c) in this case differs radically from the case of weak chaos (Fig. 50c): 1) the dependence $\tau_\hbar(N)$ is close to the power law in the case of regular semiclassical dynamics; 2) for a relatively small values of N ($N \sim 10^3$ the time-scale $\tau_\hbar$ is rather large: $\tau_\hbar \sim 30$.

14.6 Conclusion

The results presented in section 14 show that even in the case of weak chaos there appears a characteristic time-scale $\tau_\hbar \sim \ln N$ at which quantum effects violate the semiclassical approximation for the system with the Hamiltonian (14.5) (see also (12.13)). This result has a direct connection with $\ln N$ time-scale result derived in [105] for the completely integrable quantum Dicke model (14.2), when initially populated in the inverted atomic state. In this case the time-scale $\tau_\hbar \sim \ln N$ appears as a result of purely quantum fluctuations which are absent in the corresponding classical limit. Remarkably, even this integrable case shows exponential growth of quantum correlations (!) which takes place not only when the atomic sub-system is completely inverted, but also for initial population of a finite wave packet in the vicinity of the hyperbolic point. This effect of exponential enhancement of quantum correlations also takes place in the case of weak chaos. Thus, when comparing the quantum dynamics with its corresponding classically (semiclassically) unstable counterpart, the fact of instability is important, but the nature of this instability (unstable point or local instability in the case of chaotic dynamics) plays a minor role.

The time-scale $\tau_\hbar$ of violation of the semiclassical approximation has a crossover when varying the width of the initial wave packet. This "crossover-effect" for the time-scale $\tau_\hbar$ is much the same as that considered in section 9.7 for the quantum version of stochastic web. Namely, for rather wide initial wave packets populated in the stochastic layer, the classical approximation remains valid for quite large times. The experimental observation of this "crossover-effect" in the case of weak chaos would be of significant interest.

15 Quantum Dynamics in Stationary Coherent States (SCS)

15.1 Introduction

In the previous sections of this review the main statement that was discussed could be formulated in the following way. For quantum ($\hbar \neq 0$) nonlinear ($\bar{\mu} \neq 0$) systems a *finite* time-scale $\tau_\hbar$ exists, such that during the time-interval $0 < \tau < \tau_\hbar$ quantum dynamics can be well approximated by its classical counterpart (regular or chaotic). For times $\tau > \tau_\hbar$ quantum effects play a nonnegligible role.

The problem considered in this section is the following. Are there quantum nonlinear systems (Hamiltonians) for which the time-scale $\tau_\hbar = \infty$? This problem can also be formulated in a different way: Is it possible to "embed" a given classical dynamics (for example, very complicated, chaotic, and even strange attractor behavior) in quantum mechanics ? What kind of conditions on quantum Hamiltonians would we need for this ? This problem is a part of problem that has recently been widely discussed (see, for example, [109,110] and references therein). Here we would like to note that there exists a class of quantum nonlinear Hamiltonians that produce dynamics for quantum expectation values which is completely classical, and can be rather complicated (some specific examples appear in [49,111]). So, in this case $\tau_\hbar = \infty$. Although, these Hamiltonians may look rather artificial, we consider them in this review, because these Hamiltonians may be useful in constructing a bridge between quantum and classical approaches. Quantum states in which classical dynamics may be realized are first considered in [112-115], and are called stationary coherent states (SCSs). Below we consider quantum Hamiltonians which have SCSs, and present some examples, in which the dynamics for quantum expectation values in these states is chaotic (and even of a certain strange attractor type) in the sense of dynamical chaos in classical systems.

15.2 Stationary Coherent States. Basic Equations

As is well known, a coherent state (CS) can be defined as the eigenstate of an annihilation operator $\mathbf{a}$ (see (2.24)), at an initial moment of time $t = 0$. In the Heisenberg representation all operators depend on time, $\mathbf{a} = \mathbf{a}(t)$, and the wave function is time-independent.

According to [115], the state $|\alpha>$ is called a SCS, if the following condition is satisfied

$$\mathbf{a}(t)|\alpha>= \alpha(t)|\alpha>, \tag{15.1}$$

for all $t \geq 0$. In (15.1) $\mathbf{a}(t)$ is a Heisenberg operator, and $\alpha(t)$ is a c-number function. Condition (15.1) severely restricts the class of appropriate Hamiltonians. However, there does exist a class of Hamiltonians for which equation (15.1) can be satisfied. This class even includes systems with chaotic behavior in some parameter regions. In this case the dynamical chaos is completely analogous to the dynamical chaos in the classical systems.

Now we come to the basic equations for SCSs. Following [115], we consider some properties of SCSs which will be used below. In the Heisenberg representation we have the equation for $\mathbf{a}(t)$

$$i\hbar\dot{\mathbf{a}}(t) = \left[\mathbf{a}(t), \mathbf{H}\left(t, \mathbf{a}(t), \mathbf{a}^+(t)\right)\right] = \partial\mathbf{H}/\partial\mathbf{a}^+(t). \tag{15.2}$$

Consider the class of Hamiltonians that have SCSs, i.e., for which the equation (15.1) is satisfied. According to [115], the necessary and sufficient conditions for existence of SCSs are

$$[\mathbf{a}(t), \dot{\mathbf{a}}(t)]|\alpha>= 0, \tag{15.3}$$

where SCS $|\alpha>$ does not depend on time t, and is determined by the equation (15.1). The condition (15.3) is expressible as

$$[\mathbf{a}(t), [\mathbf{a}(t), \mathbf{H}(t)]]\,|\alpha>= \left\{\frac{\partial^2}{\partial[\mathbf{a}^+(t)]^2}\mathbf{H}\left(t, \mathbf{a}(t), \mathbf{a}^+(t)\right)\right\}|\alpha>= 0. \tag{15.4}$$

W-e represent the Hamiltonian $\mathbf{H}(t)$ in a power series in the operator $\mathbf{a}^+(t)$

$$\mathbf{H}\left(t, \mathbf{a}(t), \mathbf{a}^+(t)\right) = \mathbf{H}^{(0)}\left(t, \mathbf{a}(t)\right) + \mathbf{a}^+(t)\mathbf{H}^{(1)}\left(t, \mathbf{a}(t)\right) \tag{15.5}$$

$$+[\mathbf{a}^+(t)]^2\mathbf{H}^{(2)}\left(t, \mathbf{a}(t), \mathbf{a}^+(t)\right),$$

where

$$\mathbf{H}^{(2)}\left(t, \mathbf{a}(t), \mathbf{a}^+(t)\right) = \sum_{m=0}^{\infty}[\mathbf{a}^+(t)]^m\mathbf{h}_m\left(t, \mathbf{a}(t)\right). \tag{15.6}$$

The necessary and sufficient conditions for the existence of a SCS can be formulated as the requirement that there exists a function $\alpha(t)$, satisfying [115]

$$h_m\left(t, \mathbf{a}(t) = \alpha(t)\right) = 0, \quad m = 0, 1, ..., \tag{15.7}$$

$$i\hbar\dot{\alpha}(t) = H^{(1)}\left(t, \mathbf{a}(t) = \alpha(t)\right). \tag{15.8}$$

The necessity of the conditions (15.7) and (15.8) follows from (15.1), (15.3)- (15.6). To prove the sufficiency, we use the following state

$$|\Psi, t>= F(t)\mathbf{U}^+(t)|\alpha(t)>, \tag{15.9}$$

where $|\alpha(t)>$ is an eigenfunction of the operator $\mathbf{a}$ at $t = 0$ (see (2.24)), with eigenvalue $\alpha(t)$

$$\mathbf{a}|\alpha(t)>= \alpha(t)|\alpha(t)>, \tag{15.10}$$

and the function $\alpha(t)$ satisfies equations (15.7), (15.8). We represent the CS according to (2.33), (2.34), in the form

$$|\alpha(t)>= \mathbf{D}_s(\alpha(t), \alpha^*(t))\,|0>= \exp[\alpha(t)\mathbf{a}^+ - \alpha^*(t)\mathbf{a}]|0> . \quad (15.11)$$

In (15.9) $\mathbf{U}(t)$ is the evolution operator satisfying

$$i\hbar\dot{\mathbf{U}} = \mathbf{U}(t)\mathbf{H}(t), \quad \mathbf{H}(t) = \mathbf{U}^+(t)\mathbf{H}\mathbf{U}(t). \quad (15.12)$$

Now we show that the function in (15.9) $|\Psi, t>$ is an eigenfunction of the operator

$$\mathbf{a}(t) = \mathbf{U}^+(t)\mathbf{a}\mathbf{U}(t), \quad (15.13)$$

with eigenvalue $\alpha(t)$. From (15.9) and (15.10) it follows

$$\begin{aligned} \mathbf{a}(t)|\Psi, t> &= \mathbf{U}^+(t)\mathbf{a}\mathbf{U}(t)|\Psi, t> \quad (15.14) \\ &= \mathbf{U}^+(t)\mathbf{a}\mathbf{U}(t)F(t)\mathbf{U}^+(t)|\alpha(t)> \\ F(t)\mathbf{U}^+(t)\mathbf{a}|\alpha(t)> &= \alpha(t)|\Psi, t>. \end{aligned}$$

According to [114], the sufficiency of conditions (15.7), (15.8) for the existence of SCS is equivalent to the following. Require the conditions (15.7) and (15.8) to be satisfied, and represent the function $F(t)$ in (15.9) in the form

$$F(t) = \exp[\varphi(t)], \quad (15.15)$$

$$\varphi(t) = -\frac{i}{\hbar}\int_0^t H^{(0)}(t', \alpha(t'))\, dt' + \int_0^t \dot{\alpha}^*(t')\alpha(t'))dt'.$$

In this case the state $|\Psi, t>$ is stationary, i.e. the following equality is fulfilled

$$\frac{\partial|\Psi, t>}{\partial t} = 0, \quad (15.16)$$

which can be proved by the direct differentiation of (15.9) using the expressions (15.7) and (15.8). This is why the state (15.9) is stationary

$$|\Psi, t>= |\Psi, 0>= |\alpha(0)>= |\alpha> . \quad (15.17)$$

Using (15.17), find from (15.14) that

$$\mathbf{a}(t)|\Psi, 0>= \alpha(t)|\Psi, 0> . \quad (15.18)$$

Thus, the state $|\Psi, 0>=|\alpha>$ is a SCS. The evolution of the wave function of the SCS is described in the Schrödinger picture as follows

$$|\alpha, t>= \mathbf{U}(t)|\alpha>= \mathbf{U}(t)|\Psi, t> \tag{15.19}$$

$$= \mathbf{U}(t)F(t)\mathbf{U}^{+}(t)|\alpha(t)>= F(t)|\alpha(t)>$$

$$= F(t)\mathbf{D}_s(\alpha(t), \alpha^*(t))|0>,$$

where $\alpha(t)$ satisfies to the equations (15.7) and (15.8), and the function $F(t)$ is of the form (15.15). Using (15.8), we have for $F(t)$

$$F(t) = \exp\left\{\frac{i}{\hbar}\int_0^t [-H^{(0)}(t', \alpha(t')) + \alpha(t')H^{(1)*}(t', \alpha(t'))]dt'\right\}. \tag{15.20}$$

15.3 Dynamical Chaos in SCS

We now discuss how the evolution of a quantum system in SCS can lead to dynamical chaos. As the first example we consider the Hamiltonian [28]

$$\mathbf{H} = \hbar\omega\mathbf{a}^{+}\mathbf{a} + \mu\hbar^2[(\mathbf{a}^{+})^2\alpha(t)(\mathbf{a} - \alpha(t)) + \left(\mathbf{a}^{+} - \alpha^*(t)\right)\alpha^*(t)\mathbf{a}^2] \tag{15.21}$$

$$+\lambda\hbar^{1/2}(\mathbf{a}^{+} + \mathbf{a})f(t).$$

In (15.21) $f(t)$ is a given function of time, $\alpha(t)$ is a function to be determined, and ω, μ and λ are given parameters. Presenting (15.21) in the form (15.5) and (15.6), we have

$$\mathbf{H}^{(0)}(t, \mathbf{a}(t)) = \hbar^{1/2}\lambda\mathbf{a}(t)f(t) - \hbar^2\mu(\alpha^*(t))^2\mathbf{a}^2(t), \tag{15.22}$$

$$\mathbf{H}^{(1)}(t, \mathbf{a}(t)) = \hbar\omega\mathbf{a}(t) + \mu\hbar^2\alpha^*(t)\mathbf{a}^2(t) + \hbar^{1/2}\lambda f(t),$$

$$\mathbf{H}^{(2)}(t, \mathbf{a}(t)) = \hbar^2\mu\alpha(t)[\mathbf{a}(t) - \alpha(t)].$$

Equation (15.7) for $h_0(t, \alpha(t))$ holds identically for the Hamiltonian (15.21), (15.22), and equation (15.8) becomes

$$i\dot{\alpha}(t) = \omega\alpha(t) + \mu\hbar|\alpha(t)|^2\alpha(t) + \frac{\lambda}{\sqrt{\hbar}}f(t). \tag{15.23}$$

To simplify the analysis, we choose $f(t)$ as a periodic time sequence of δ-pulses

$$f(t) = \sum_{n=-\infty}^{\infty} \delta(t - nT_0). \tag{15.24}$$

In this case equation (15.23) becomes

$$i\frac{dz(\tau)}{d\tau} = z(\tau) + |z(\tau)|^2 + \varepsilon \sum_{n=-\infty}^{\infty} \delta(\tau - nT), \tag{15.25}$$

where we have introduced the following definitions

$$z(\tau) = (\mu\hbar/\omega)^{1/2}\alpha(t), \quad \varepsilon = \lambda(\mu/\omega)^{1/2}, \quad \tau = \omega t, \quad T = \omega T_0. \tag{15.26}$$

Equation (15.25) may be written in Hamiltonian form as

$$\frac{dz}{d\tau} = -i\frac{\partial}{\partial z^*}H(\tau, z, z^*), \quad \frac{dz^*}{d\tau} = i\frac{\partial}{\partial z}H(\tau, z, z^*), \tag{15.27}$$

where

$$H(\tau, z, z^*) = zz^* + \frac{1}{2}z^2 z^{*2} + \varepsilon(z + z^*) \sum_{n=-\infty}^{\infty} \delta(\tau - nT). \tag{15.28}$$

The criterion of stochastic behavior for the system (15.25) is derived in [116] by investigating the map

$$z_n = \exp\{-i(1 + |z_{n-1} - i\varepsilon|^2)T\}(z_{n-1} - i\varepsilon), \tag{15.29}$$

where the following variables are used

$$z_n \equiv z(\tau_n - 0), \quad \bar{z}_n \equiv z(\tau_n + 0), \quad \tau_n = nT. \tag{15.30}$$

The variables z_n and $\bar{z}_n$ are related by

$$z_n = \bar{z}_{n-1} \exp[-i(1 + |\bar{z}_{n-1}|^2 T], \quad \bar{z}_n = z_n - i\varepsilon. \tag{15.31}$$

The map (15.29) coincides with the map (6.23) under the condition: $2\mu = 1$. Thus, the condition for stochastic behavior for the map (15.29) takes, according to (15.34), the form

$$K_n \sim 2\varepsilon T I_n^{1/2} > 1. \tag{15.32}$$

Thus, if we require that the function $\alpha(t) = (\omega/\mu\hbar)^{1/2} z(\tau)$ in Hamiltonian (15.21) satisfies equation (15.25), the CS $|\alpha>$ for the system with the Hamiltonian (15.21) will be stationary (SCS). The expectation value for the operator $\mathbf{a}(\tau)$ in this SCS is given by the function $z(\tau)$

$$< \alpha|\mathbf{a}(\tau)|\alpha >= \left(\frac{\omega}{\mu\hbar}\right)^{1/2} z(\tau). \tag{15.33}$$

The behavior of the function $z(\tau)$ strongly depends on the value of the parameter of stochasticity K_n in (15.32). If $K_n < 1$, the behavior of the function $z(\tau)$ is regular in time. If $K_n > 1$, the dynamics of the function $z(\tau)$ is chaotic. The evolution of the wave function in the Schrödinger representation is described for the Hamiltonian (15.21) by the expression, which follows from (15.19), (15.20), (15.22)

$$|z, \tau > \equiv \mathbf{U}(\tau)|\Psi, 0 >= F(\tau)|z(\tau)> \tag{15.34}$$

$$= \exp\left\{\frac{i\omega}{\mu\hbar}\int_0^\tau (1 + 2|z(\tau')|^2)|z(\tau')|^2 d\tau'\right\}|z(\tau)>,$$

$$|z(\tau)>= \mathbf{D}_s\left(z(\tau), z^*(\tau)\right)|0>.$$

Fig. 2a shows a typical stochastic trajectory for the map (15.29) on the plane (I_n, θ_n), where $z_n = \sqrt{I_n}\exp(-i\theta_n)$. One can see from Fig. 2a that the boundary in action I_b between chaotic and regular dynamics agrees rather well with the estimate

$$I_b \approx \frac{1}{2\varepsilon T} = \frac{1}{4},$$

which follows from the condition $K_b \approx 1$. There are some differences from the value I_b, which can be attributed to the first term in (15.34) ($\varepsilon/I_b^{1/2} = 0.2$), and the finite trajectory length. Fig. 2b shows the time dependence of the logarithm of the distance $\rho(\tau) = |z(\tau) - z'(\tau)|$ between the trajectory $z(\tau)$, and the trajectory $z'(\tau)$, which is close to the trajectory $z(\tau)$ at $\tau = 0$. One can see from Fig. 2b, that the trajectory $z(\tau)$ is locally unstable. This is a characteristic property of the stochastic regime of motion in classical dynamical systems.

15.4 Stability of SCS

Since the map (15.29) can generate unstable motion, the question arises whether such an instability of the function z_n results the instability of the SCS (15.34). In other words, whether the SCS (15.34) will be destroyed in this case. The answer to this question is the following. The SCS is stable in the sense of violation of wave packet (the Schrödinger wave function (15.34)), even in the case when the map (15.29) generates a stochastic trajectory. The proof of this statement is as follows. Condition (15.29) implies that equation (15.8) holds during the time interval between between two kicks. Let the Schrödinger wave function at $\tau = n$ be a SCS. Then, according to (15.34), it has the form

$$|z, n >= F(n)|z_n >, \tag{15.35}$$

and satisfies the condition

$$\mathbf{a}|z, n >= z_n|z, n > . \tag{15.36}$$

Assuming z_n to be known, and using (15.29), we construct the number z_{n+1}, and, then, the coherent state $|z_{n+1} >$. Hence, we have from (15.34)

$$F(n+1) = F(n) + \int_n^{n+1} (\cdots) d\tau.$$

After this, the function $|z, n+1 >$ is constructed, which is the coherent state, satisfying the condition

$$\mathbf{a}|z, n+1 >= z_{n+1}|z, n+1 > . \tag{15.37}$$

Thus, the state $|z, n >$ is a SCS for arbitrary n. Thus, errors in calculating the function z_n using the map (15.29) do not result in violation of the SCS property, but merely result in errors in the c-number function z_n, which is the usual property of trajectories undergoing chaotic motion.

15.5 Diffusion in SCS

Consider the dynamics of the average number of quanta in the SCS (15.34). The probability amplitude to find n quanta at time τ in a

SCS has, according to (15.37), the form

$$< n|z,\tau >= F(\tau) < n|z(\tau) >= F(\tau)\frac{z^n(\tau)}{\sqrt{n!}}e^{-|z(\tau)|^2/2}. \qquad (15.38)$$

Thus, the distribution function of quanta in state $|z,\tau>$ at time τ has, according to (15.38), the Poissonian form

$$P[z(n)] \equiv |< n|z(\tau) >|^2 = e^{-n(\tau)}\frac{\bar{n}^n(\tau)}{n!}, \quad (\bar{n}(\tau) \equiv |z(\tau)|^2). \qquad (15.39)$$

According to (15.29), in the stochastic regime the average number of quanta $\bar{n}(\tau)$ in SCS grows in time nearly according to the diffusion law

$$I_\tau \equiv \bar{n}(\tau) = I_0 + \varepsilon^2\tau. \qquad (15.40)$$

15.6 Strange Attractor in SCS

In this section we discuss an example which leads to a strange attractor in the SCS [29]. It will be shown that, although the initial Hamiltonian (15.5) is Hermitian, the dynamics of its quantum expectation values may be determined by trajectories on a set of strange attractor type.

Consider Hermitian Hamiltonian

$$\mathbf{H} = \hbar\omega\mathbf{a}^+\mathbf{a} + \mu\hbar^2[(\mathbf{a}^+)^2\alpha(t)\left(\mathbf{a}-\alpha(t)\right) + \left(\mathbf{a}^+ - \alpha^*(t)\right)\alpha^*(t)\mathbf{a}^2] \qquad (15.41)$$
$$+\hbar\nu\{\mathbf{a}^+[-i\alpha(t) + i[\alpha^*(t)]\mathbf{a}\} + \lambda\hbar^{1/2}(\mathbf{a}^+ + \mathbf{a})f(t).$$

From (15.41) we have for the operators $\mathbf{H}^{(0)}$, $\mathbf{H}^{(1)}$, and $\mathbf{H}^{(2)}$ (see (15.5), (15.6))

$$\mathbf{H}^{(0)}(t,\mathbf{a}(t)) = \hbar^{1/2}\lambda\mathbf{a}(t)f(t) - \hbar^2\mu[\alpha^*(t)]^2\mathbf{a}^2(t) + i\hbar\nu\alpha^*(t)\mathbf{a}(t), \qquad (15.42)$$
$$\mathbf{H}^{(1)}(t,\mathbf{a}(t)) = \hbar\omega\mathbf{a}(t) + \mu\hbar^2\alpha^*(t)\mathbf{a}^2(t) - i\hbar\nu\alpha(t) + \hbar^{1/2}\lambda f(t),$$
$$\mathbf{H}^{(2)}\left(t,\mathbf{a}(t),\mathbf{a}^+(t)\right) = \hbar^2\mu\alpha(t)[\mathbf{a}(t) - \alpha(t)],$$

where ω, μ, λ and ν are real parameters; the function $f(t)$ is determined in (15.24); and $\alpha(t)$ is a function to be determined below.

In the case of Hamiltonian (15.41) the equation (15.8) takes the form

$$i\frac{dz(\tau)}{d\tau} = z(\tau) + |z(\tau)|^2 z(\tau) - \frac{i}{2}\gamma z(\tau) + \varepsilon \sum_{n=-\infty}^{\infty} \delta(\tau - nT), \tag{15.43}$$

with the notations introduced in (15.26), and

$$\gamma = 2\nu/\omega. \tag{15.44}$$

Equation (15.7) is always satisfied. Thus, if the function $z(\tau)$ satisfies equation (15.46), then the coherent state $|\alpha>$ (15.1) is a SCS.

We now dwell on some properties of equation (15.43), and show how solutions of strange attractor type may appear in this state. At $\varepsilon = 0$ the solution of equation (15.43) has the form

$$|z(\tau)|^2 = |z(0)|^2 e^{-\gamma\tau}. \tag{15.45}$$

When $\varepsilon \neq 0$, integration of the equation (15.43) from $\tau_n \equiv nT - 0$ to $\tau_{n+1} = (n+1)T - 0$ leads to the following map

$$z_{n+1} = \bar{\gamma}(z_n - i\varepsilon)\exp[-i(1 + q|z_n - i\varepsilon|^2)T], \tag{15.46}$$

where

$$z_n \equiv z(nT - 0), \quad \bar{\gamma} = \exp(-\gamma T/2), \tag{15.47}$$

$$q = \frac{1 - \exp(-\bar{\gamma}T)}{\bar{\gamma}T} = \frac{1 - \bar{\gamma}^2}{\bar{\gamma}T}.$$

Introducing the variables I_n and θ_n in the same way as before, we have from (15.46) the map

$$I_{n+1} = \bar{\gamma}^2(I_n + 2\varepsilon\sqrt{I_n}\sin\theta_n + \varepsilon^2), \tag{15.48}$$

$$\theta_{n+1} = \arctan\left(\tan\theta_n + \frac{\varepsilon}{\sqrt{I_n}\cos\theta_n}\right) + T + \frac{qT}{\bar{\gamma}^2}I_{n+1}.$$

The parameter of stochasticity for the map (15.48) has the form

$$\frac{\partial\theta_{n+1}}{\partial\theta_n} = \frac{\left(1 + \frac{\varepsilon}{I_n^{1/2}}\sin\theta_n\right)}{\left(1 + \frac{\varepsilon^2}{I_n} + \frac{2\varepsilon}{I_n^{1/2}}\sin\theta_n\right)} + \mathcal{K}_n\cos\theta_n, \tag{15.49}$$

where

$$\mathcal{K}_n = qK_n = 2\varepsilon T\sqrt{I_n q}. \tag{15.50}$$

When $\varepsilon/\sqrt{I_n} \ll 1$, a rough estimate for the criterion of the appearance of a strange attractor can be written, in analogy with [117], as follows:

$$\mathcal{K}_n > 1, \quad \left(\sqrt{I_n} > \frac{1}{2\varepsilon T q}\right).$$

In this case, the motion of the system appears to be locally unstable. A detailed analysis of the map of the type (15.48), leading to the strange attractor, is given in [117] (the map considered in [117] follows from (15.41) if one puts $\omega = 0$ in (15.41), and uses an appropriate substitution of variables).

We make two remarks to clear up the properties of the SCS. The first one concerns equation (15.1). In the general case, the solution of the operator equation for $\mathbf{a}(t)$ (15.2) has the form $\mathbf{a}(t) = \mathbf{a}(t, \mathbf{a}, \mathbf{a}^+)$. Then, because of the presence of the operator $\mathbf{a}^+$, satisfying

equation (15.1) seems to be impossible. However, as one can see, in the case (15.21) or (15.41), there are the following equations

$$\frac{d^n\mathbf{a}(t)}{dt^n}|_{t=0}|\alpha>= g_n|\alpha>, \quad n = 1, 2, ..., \tag{15.51}$$

where g_n are c-number functions. Construct the operator

$$\mathbf{a}(t) = \sum_{n=0}^{\infty} \frac{t^n}{n!}\frac{d^n\mathbf{a}(t)}{dt^n}|_{t=0}. \tag{15.52}$$

Equation (15.1) then follows from (15.51) and (15.52).

The second remark is connected with the equation (15.43), which includes the dissipative term $-i(\gamma/2)z$, although Hamiltonian (15.41) is Hermitian. To explain this we give a simple classical example. Consider the following classical Hamiltonian

$$H_{cl}(z^*, z, \tau) = z^*z + \{z^{*2}\eta(\tau)[z - \eta(\tau)] + [z^* - \eta^*(\tau)]\eta^*(\tau)z^2\} \tag{15.53}$$

$$+\frac{\gamma}{2}\{z^*[-i\eta(\tau)] + [i\eta^*(\tau)]z\} + \varepsilon(z^* + z)\sum_{n=-\infty}^{\infty}\delta(\tau - nT),$$

where $\eta(\tau)$ is some function determined below by an additional condition. In (15.53) the variables z^* and z are considered as canonical. Thus, the Hamiltonial equations have the form

$$i\dot{z} = \partial H_{cl}/\partial z^*, \quad i\dot{z}^* = -\partial H_{cl}/\partial z. \tag{15.54}$$

We shall solve equations (15.54) with an additional condition on the function $\eta(\tau)$, namely

$$\eta(\tau) = z(\tau). \tag{15.55}$$

One can see that the canonical equations (15.54) along with the condition (15.55) lead to equation (15.43). In this example, the equations (15.54) are analogous to the quantum Heisenberg equations, and the condition (15.55) is analogous to the condition for the coherent state to be a SCS.

In conclusion we note, that the examples of SCSs with Hamiltonians (15.21) and (15.41) show the possibility of the appearance of chaotic motion (even of a strange attractor type) in quantum dynamical systems. For these Hamiltonians the dynamics of quantum expectation values is determined entirely by classical equations, and in this case $\tau_\hbar = \infty$. The physical meaning of such Hamiltonians, and their connection with real physical situations has not been deeply studied yet. Nevertheless, these Hamiltonians may could be interesting, since they represent a class of quantum dynamical systems that produce classical dynamics for expectation values.

16 Discussion

The main problem discussed in this review is connected with the classical/quantum crossover time-scale $\tau_\hbar$ (we call it the "$\tau_\hbar$-problem"), which occurs when considering dynamics of quantum boson and spin systems, in the quasiclassical region of parameters. There are several aspects of this problem. One aspect concerns the fundamental problem of correspondence between quantum and classical mechanics. This problem has existed sinse the creation of quantum mechanics, and has been widely discussed in the scientific literature. Recently this problem has attracted additional interest, because of intensive investigation of a new class of classical dynamical systems which exhibit chaotic behavior in some regions of phase space. In this case classical dynamical solutions are unstable in the sense of local (exponential) instability of classical trajectories in phase space, and quantum effects can play a significant role even in the quasiclassical region. Now, classical chaos occurs in any nonintegrable classical dynamical system. Nonintegrability means that the number of global independent integrals of motion in the system under consideration is less then the number of degrees of freedom, and no separation of variables occurs in either the classical Newton's equations, or in the corresponding Schödinger equation. We would like to note here, that according to this definition, "quantum nonintegrability" occurs just as well in dynamical systems with either discrete, or continuous energy (quasienergy) spectra.

The $\tau_\hbar$-problem has recently acquired important potential applications for mesoscopic systems in the quickly developing field of nanotechnology, since these systems are both nonintegrable and quasiclassical.

In this review we investigate the $\tau_\hbar$-problem by directly comparing quantum $X_i(t)$ and corresponding classical $X_i^{(cl)}(t)$ $(i = 1, 2, ...)$ time-dependent expectation values. It is well known that according to the correspondence principle the following condition is satisfied

$$\lim_{\hbar \to 0} X_i(t) = X_i^{(cl)}(t). \tag{16.1}$$

The problem with this limit is that in reality $\hbar \neq 0$! In this review we consider the $\tau_\hbar$-problem by using a *direct* method based on the exact c-number equations for quantum time-dependent expectation values in spin and boson coherent states. An the additional procedure of averaging over the initial density matrix is also used (see sections 5.2, 12.7) for initial conditions that are not coherent states. From our point of view the CS method has some advantages over density matrix methods, which do not have a unique representation (see section 10.2), and often lead to a mixture of "mathematical and physical effects", especially for unstable classical solutions.

We show for quantum dynamical systems with nonlinear Hamiltonians (i.e., Hamiltonians with non-equidistant spectrum), that the characteristic time-scale $\tau_\hbar$ exists, and for $0 < \tau < \tau_\hbar$ the quantum dynamics for expectation values coincides approximately with the corresponding classical dynamics. For $\tau > \tau_\hbar$ quantum effects are nonnegligible. Thus, at $\tau \sim \tau_\hbar$ a "classical-quantum crossover" takes place in the dynamics of expectation values and correlation functions.

The time-scale $\tau_\hbar$ is usually different for integrable and nonintegrable quantum systems. For integrable quantum systems, in stable regions of motion, $\tau_\hbar \sim 1/\hbar^\alpha$, where $\alpha \sim 1$. For nonintegrable (classically chaotic) systems $\tau_\hbar \sim C_1 \ln(C_2/\hbar)$, a time which in the quasiclassical region is much shorter than for integrable systems. The problem arises of measuring this crossover-time $\tau_\hbar$ experimentally for classically chaotic systems. This problem is rather important, since the time-scale $\tau_\hbar$ is the smallest time-scale that appears in comparing quantum and classical Hamiltonian dynamics. To succeed in measuring $\tau_\hbar$ experimentally, one must choose an appropriate system in the quasiclassical region of parameters.

We suggest in this review some models which we believe are promising for this aim. These models are different nonintegrable generalizations of the quantum Dicke model [24] well known in nonlinear optics which represents an ensemble of N two-level atoms placed in a high quality resonator, and interacting with a single resonant eigenmode. In the resonant approximation, considered in [24], this system is completely integrable. In this model the semiclassical limit corresponds to considering the radiation field in the classical approximation.

We discuss in this review two variants of Dicke model that produce semiclassical chaotic dynamics. The first variant accounts for nonresonant terms [94], and the second one [102-104] imposes an additional external coherent field, whose frequency slightly differs from atomic transition frequency. In the latter case, the resonant approximation can be still used, and the transition to strong chaos occurs at rather small values of the constants characterizing the interaction of atoms with the radiation and with the external fields. We also consider a similar system, which is a magnetic variant of the last one: an ensemble of N two-level paramagnetic spins in a constant magnetic field, placed into a resonator and interacting with one resonant eigenmode of the resonator, and with an external resonant field. Also, the frequency of the spin's precession in a constant magnetic field is assumed to be slightly different from the frequency of the external resonant field.

We quantize the resonator eigenmode, and consider all these systems as pure quantum dynamics. The main results which are discussed in this part of review are the following. Under the condition of dynamical chaos in semiclassical approximation, the time-scale $\tau_\hbar \sim \ln N$. Also we show that the quantum correlation functions grow exponentially in this case during the times $0 < \tau < \tau_\hbar$. We hope, that this exponential growth of quantum correlations in systems like these will eventually be experimentally observed.

Note that the problem considered here is directly related to the validity of the well known $1/N$ expension (see, for example, the review [118]).

We also consider in this review a special class of quantum nonlinear Hamiltonians which admit the so-called stationary coherent states. In these states $\tau_\hbar = \infty$, and the dynamics for expectation values may be chaotic, and even of a strange attractor type. These Hamiltonians are rather artificial from the viewpoint of quantum mechanics, but may be useful in constructing a bridge between classical and quantum mechanics for large systems.

Appendix A

addtocontentstoc 1. We prove here only the last formula in (2.47).

Using (2.34), (2.36), we have

$$Q \equiv < z|\mathbf{af}|z >= e^{-|z|^2} < 0|e^{-z\mathbf{a}^+}|e^{z^*\mathbf{a}}\mathbf{af}e^{z\mathbf{a}^+}|e^{-z^*\mathbf{a}}|0 > . \qquad (A.1)$$

Now, using the trivial property of the annihilation operator

$$e^{-z^*\mathbf{a}}|0 >= |0 >, \qquad (A.2)$$

we derive from (A.1)

$$Q = e^{-|z|^2}\frac{\partial}{\partial z^*} < 0|e^{z^*\mathbf{a}}\mathbf{f}e^{z\mathbf{a}^+}|0 >= e^{-|z|^2}\frac{\partial}{\partial z}e^{|z|^2}f. \qquad (A.3)$$

All other formulas in (2.47) can be derived analogously.

2. Here we prove formulas (2.65). First, calculate the expression $< \xi|\mathbf{fS}^+|\xi >$. From (2.51) we have

$$\frac{\partial}{\partial\xi}|\xi >= \mathbf{S}^+|\xi > -\frac{S\xi^*}{(1+|\xi|^2)}|\xi >, \qquad (A.4)$$

$$\frac{\partial}{\partial\xi} < \xi| = -\frac{S\xi^*}{(1+|\xi|^2)} < \xi|. \qquad (A.5)$$

From (A.4), (A.5) we derive for $\partial f/\partial\xi$

$$\frac{\partial f}{\partial\xi} \equiv \frac{\partial}{\partial\xi} < \xi|\mathbf{f}|\xi >= \left(\frac{\partial}{\partial\xi} < \xi|\right)\mathbf{f}|\xi > \qquad (A.6)$$

$$+ < \xi|\mathbf{f}\left(\frac{\partial}{\partial\xi}|\xi >\right) = -\frac{2S\xi^*}{(1+|\xi|^2)}f + < \xi|\mathbf{fS}^+|\xi > .$$

Then, we have from (A.6)

$$< \xi|\mathbf{fS}^+|\xi >= \frac{\partial f}{\partial\xi} + \frac{2S\xi^*}{(1+|\xi|^2)}f = (1+|\xi|^2)^{-2S}\frac{\partial}{\partial\xi}(1+|\xi|^2)^{2S}f. \quad (A.7)$$

To calculate the expression $< \xi|\mathbf{f}\mathbf{S}^z|\xi >$, note that

$$\mathbf{S}^z = \mathbf{S}^+\xi + e^{\xi \mathbf{S}^+}\mathbf{S}^z e^{-\xi \mathbf{S}^+}, \tag{A.8}$$

which follows from the operator equality

$$\mathbf{f}_1(\xi) \equiv e^{-\xi \mathbf{S}^+}\mathbf{S}^z e^{\xi \mathbf{S}^+} = \mathbf{S}^+\xi + \mathbf{S}^z. \tag{A.9}$$

The operator expression (A.9) can be easily proved. Using the commutation relations (2.49) we have the differential equation for $\mathbf{f}_1(\xi)$

$$\frac{d\mathbf{f}_1(\xi)}{d\xi} = \mathbf{S}^+, \tag{A.10}$$

which leads directly to the right side of (A.9). Then, using the definition of spin CS (2.51), it follows from (A.8) that

$$< \xi|\mathbf{f}\mathbf{S}^z|\xi >= \xi < \xi|\mathbf{f}\mathbf{S}^+|\xi > -Sf. \tag{A.11}$$

Using (A.7), we derive from (A.11) the corresponding formula in (2.65). Now we prove the formula for $< \xi|\mathbf{S}^-\mathbf{f}|\xi >$. We have from the definition of spin CS

$$\frac{\partial}{\partial \xi^*} < \xi| = -\frac{S\xi}{(1+|\xi|^2)} < \xi| + < \xi|\mathbf{S}^-, \tag{A.12}$$

$$\frac{\partial}{\partial \xi^*}|\xi >= -\frac{S\xi}{(1+|\xi|^2)}|\xi >.$$

From (A.12) we derive analogously to (A.6)

$$\frac{\partial f}{\partial \xi^*} = -\frac{2S\xi}{(1+|\xi|^2)}f + < \xi|\mathbf{S}^-\mathbf{f}|\xi > . \tag{A.13}$$

From (A.13) we derive the formula

$$< \xi|\mathbf{S}^-\mathbf{f}|\xi >= \frac{\partial f}{\partial \xi^*} + \frac{2S\xi}{(1+|\xi|^2)}f = (1+|\xi|^2)^{-2S}\frac{\partial}{\partial \xi^*}(1+|\xi|^2)^{2S}f. \tag{A.14}$$

Now we derive the formula for $<\xi|\mathbf{f}\mathbf{S}^-|\xi>$. For this we use the the following operator

$$\mathbf{f}_2(\xi) \equiv e^{-\xi\mathbf{S}^+}\mathbf{S}^- e^{\xi\mathbf{S}^+}, \tag{A.15}$$

The operator $\mathbf{f}_2$ satisfies the following equation

$$\frac{d\mathbf{f}_2(\xi)}{d\xi} = -2\mathbf{f}_1(\xi), \tag{A.16}$$

where the operator $\mathbf{f}_1$ is determined in (A.9). Substituting (A.9) in (A.16), and taking into account the initial condition $\mathbf{f}_2(0) = \mathbf{S}^-$ gives

$$\mathbf{f}_2(\xi) = -\mathbf{S}^+\xi^2 - 2\mathbf{S}^z\xi + \mathbf{S}^-. \tag{A.17}$$

Then, we have from (A.15), (A.17)

$$\mathbf{S}^- e^{\xi\mathbf{S}^+} = e^{\xi\mathbf{S}^+}(-\mathbf{S}^+\xi^2 - 2\mathbf{S}^z\xi + \mathbf{S}^-). \tag{A.18}$$

Using (A.7) we derive from (A.18) the corresponding formula in (2.65). To prove the formula $<\xi|\mathbf{S}^+\mathbf{f}|\xi>$, one should use the Hermitian conjugates of expressions (A.18) and (A.14). Finally, the formula for $<\xi|\mathbf{S}^z\mathbf{f}|\xi>$ may be derived by using the expressions (A.9) and (A.14).

3. Here we prove formula (3.8). We have

$$<\alpha|\mathbf{a}^n(\mathbf{a}^+)^k|\alpha> = e^{-|\alpha|^2} <0|e^{\alpha^*\mathbf{a}}\mathbf{a}^n(\mathbf{a}^+)^k e^{\alpha\mathbf{a}^+}|0> \tag{A.19}$$

$$= e^{-|\alpha|^2}\frac{\partial^n}{\partial(\alpha^*)^n} <0|e^{\alpha^*\mathbf{a}}(\mathbf{a}^+)^k e^{\alpha\mathbf{a}^+}|0>$$

$$= e^{-|\alpha|^2}\frac{\partial^n}{\partial(\alpha^*)^n} <\alpha|(\mathbf{a}^+)^k|\alpha> e^{|\alpha|^2}$$

$$= e^{-|\alpha|^2}\frac{\partial^n}{\partial(\alpha^*)^n}(\alpha^*)^k e^{|\alpha|^2} = e^{-|\alpha|^2}\frac{\partial^k}{\partial(\alpha^*)^k}\alpha^n e^{|\alpha|^2}.$$

4. We prove now formula (3.33). For this we express (3.33) in an equivalent form

$$H exp\left(\frac{\overleftarrow{\partial}}{\partial\alpha}\frac{\overrightarrow{\partial}}{\partial\alpha^*}\right) f = H\left(\alpha^*, \alpha + \frac{\overrightarrow{\partial}}{\partial\alpha^*}\right) f, \tag{A.20}$$

$$f exp\left(\frac{\overleftarrow{\partial}}{\partial\alpha}\frac{\overrightarrow{\partial}}{\partial\alpha^*}\right) H = f H\left(\alpha^* + \frac{\overleftarrow{\partial}}{\partial\alpha}, \alpha\right) \tag{A.21}$$

$$= H\left(\alpha^* + \frac{\overrightarrow{\partial}}{\partial\alpha}, \alpha\right) f.$$

Let H has the form (3.34). Using (3.34), we have from (A.20) the following expression

$$H exp\left(\frac{\overleftarrow{\partial}}{\partial\alpha}\frac{\overrightarrow{\partial}}{\partial\alpha^*}\right) f = \sum_{m,n} H_{m,n}(\alpha^*)^m \left(\alpha + \frac{\partial}{\partial\alpha^*}\right)^n f. \tag{A.22}$$

Then, we use the relations

$$\left(\alpha + \frac{\partial}{\partial\alpha^*}\right)^n f = \sum_{k=0}^{n} C_k^n \alpha^k \left(\frac{\partial}{\partial\alpha^*}\right)^{n-k} f, \tag{A.23}$$

$$\alpha^k = e^{-|\alpha|^2} \frac{\partial^k}{\partial(\alpha^*)^k} e^{|\alpha|^2},$$

to derive the following expression

$$\left(\alpha + \frac{\partial}{\partial\alpha^*}\right)^n = e^{-|\alpha|^2} \sum_{k=0}^{n} C_k^n \left\{\left(\frac{\partial}{\partial\alpha^*}\right)^k e^{|\alpha|^2}\right\} \left(\frac{\partial}{\partial\alpha^*}\right)^{n-k} f \tag{A.24}$$

$$= e^{-|\alpha|^2}\left(\frac{\partial}{\partial\alpha^*}\right)^n f e^{|\alpha|^2}.$$

Analogously, we have

$$\left(\alpha^* + \frac{\partial}{\partial\alpha}\right)^n f = e^{-|\alpha|^2} \frac{\partial^n}{\partial\alpha^n} f e^{|\alpha|^2}. \tag{A.25}$$

Using (A.24), we derive from (A.22)

$$H exp\left(\frac{\overleftarrow{\partial}}{\partial\alpha}\frac{\overrightarrow{\partial}}{\partial\alpha^*}\right) f = e^{-|\alpha|^2} \sum_{m,n} (\alpha^*)^m \left(\frac{\partial}{\partial\alpha^*}\right)^n f e^{|\alpha|^2} \tag{A.26}$$

$$= e^{-|\alpha|^2} H\left(\alpha^*, \frac{\partial}{\partial\alpha^*}\right) f e^{|\alpha|^2},$$

which corresponds to the first term in the operator (3.12). Analogously, we get the second term in (3.12).

Appendix B

From (9.18) and (9.8) we have

$$\mathcal{R}_n(\gamma|\beta) =< \Psi_{n-1}|exp\{i\kappa\cos[(k/\sqrt{2})(\mathbf{a}^+ + \mathbf{a})]\}(\gamma e^{iT}\mathbf{a}^+) \quad (B.1)$$
$$\times exp(\beta e^{-iT}\mathbf{a})exp\{-i\kappa\cos[(k/\sqrt{2})(\mathbf{a}^+ + \mathbf{a})]\}|\Psi_{n-1}>,$$

where the ordering [33]

$$e^{\tau\mathbf{a}^+\mathbf{a}}f(\mathbf{a},\mathbf{a}^+)e^{-\tau\mathbf{a}^+\mathbf{a}} = f(\mathbf{a}e^{-\tau},\mathbf{a}^+e^{\tau}) \quad (B.2)$$

is used. To make (B.1) simpler, let us do the following. Write (B.1) as

$$\mathcal{R}_n(\gamma|\beta) =< \Psi_{n-1}|exp(\gamma e^{iT}\mathbf{a}^+)exp(-\gamma e^{iT}\mathbf{a}^+) \quad (B.3)$$
$$exp\{i\kappa\cos[(k/\sqrt{2})(\mathbf{a}^+ + \mathbf{a})]\}$$
$$\times exp(\gamma e^{iT}\mathbf{a}^+)exp(-\gamma e^{iT}\mathbf{a}^+)exp\{-i\kappa\cos[(k/\sqrt{2})(\mathbf{a}^+ + \mathbf{a})]\}$$
$$\times exp(-\beta e^{-iT}\mathbf{a})exp(\beta e^{-iT}\mathbf{a})|\Psi_{n-1}>,$$

and use the ordering [33]

$$e^{-\tau\mathbf{a}^+}f(\mathbf{a},\mathbf{a}^+)e^{\tau\mathbf{a}^+} = f(\mathbf{a}+\tau,\mathbf{a}^+), \quad (B.4)$$
$$e^{\tau\mathbf{a}}f(\mathbf{a},\mathbf{a}^+)e^{-\tau\mathbf{a}} = f(\mathbf{a},\mathbf{a}^+ + \tau).$$

Taking account of (B.4) we obtain from (B.3)

$$\mathcal{R}_n(\gamma|\beta) =< \Psi_{n-1}|exp(\gamma e^{iT}\mathbf{a}^+)exp\{i\kappa\cos[(k/\sqrt{2})(\mathbf{a}^+ + \mathbf{a} + \gamma e^{iT})]\} \quad (B.5)$$
$$\times exp\{-i\kappa\cos[(k/\sqrt{2})(\mathbf{a}^+ + \mathbf{a} + \beta e^{-iT})]\}exp(\beta e^{-iT})|\Psi_{n-1}>.$$

A significant point for further simplification is the fact that functions in (B.5) having κ, commute, since the arguments of the corresponding cosines are different only in the c-number. This makes it possible to collect the exponents having κ in (B.5) and to represent $\mathcal{R}_n(\gamma|\beta)$ in the form

$$\mathcal{R}_n(\gamma|\beta) =< \Psi_{n-1}|exp(\gamma e^{iT}\mathbf{a}^+)exp\{2i\kappa\sin[(k/\sqrt{2})(\beta e^{-iT} - \gamma e^{iT})] \quad (B.6)$$

$$\times \sin\left[(k/\sqrt{2})\left(\mathbf{a}^+ + \mathbf{a} + \frac{\gamma e^{iT} + \beta e^{-iT}}{2}\right)\right]\} exp(\beta e^{-iT}\mathbf{a})|\Psi_{n-1}>$$

$$= \sum_{m=-\infty}^{\infty} J_m\{2\kappa \sin[(k/\sqrt{2})(\gamma e^{iT} + \beta e^{-iT})]\}$$

$$\times < \Psi_{n-1}|exp(\gamma e^{iT}\mathbf{a}^+)exp[im(k/\sqrt{2})(\mathbf{a}^+ + \mathbf{a})]exp(\beta e^{-iT}\mathbf{a})|\Psi_{n-1}> .$$

The average $< \Psi_{n-1}| \cdots |\Psi_{n-1}>$ entering (B.6) can be expressed as

$$< \Psi_{n-1}|\cdots|\Psi_{n-1} >= exp(-m^2k^2/4) < \Psi_{n-1}|exp[(\gamma e^{iT} + imk/\sqrt{2})\mathbf{a}^+]$$

$$\times exp[(\beta e^{-iT} + imk/\sqrt{2})\mathbf{a}]|\Psi_{n-1}> \qquad (B.7)$$

$$= exp(-m^2k^2/4)\mathcal{R}_{n-1}(\gamma e^{iT} + imk/\sqrt{2}|\beta e^{-iT} + imk/\sqrt{2}).$$

Substituting (B.7) into (B.6) we obtain the recursion relation (9.19).

Appendix C

1. To prove the formula (10.29), we consider the operator

$$\mathbf{B}(\theta, \hat{n}) = \sum_p \frac{1}{2\pi} \int d\varphi a(\varphi, p)\hat{\Delta}(\varphi, p), \qquad (C.1)$$

where $a(\varphi, p)$ and $\hat{\Delta}(\varphi, p)$ are determined by formulas (10.27) and (10.28), respectively. In (C.1) and below integration is implied on the segment $[0, 2\pi]$, and summation from $-\infty$ to ∞.

To prove that $\mathbf{B}(\theta, \hat{n}) \equiv \mathbf{A}(\theta, \hat{n})$ it is enough to show that the matrix elements of these operators are equal in an arbitrary complete set of functions. Take as a normalized complete set, the set of functions

$$|s>= \frac{1}{\sqrt{2\pi}} e^{is\theta}, \quad (s = 0, \pm 1, \ldots). \qquad (C.2)$$

Substituting (10.27) in (C.1), we have for the matrix element of the operator $\mathbf{B}$

$$< s|\mathbf{B}|q >\equiv \frac{1}{2\pi} \int e^{-is\theta}\mathbf{B}(\theta, \hat{n})e^{iq\theta} d\theta \qquad (C.3)$$

$$= \sum_p \frac{1}{2\pi} \int d\phi \left[\sum_{l,r} < l|\mathbf{A}|r >< r|\hat{\Delta}(\varphi, p)|l >\right] < s|\hat{\Delta}(\varphi, p)|q > .$$

We show the validity of the following formula

$$\sum_p \frac{1}{2\pi}\int d\varphi < r|\hat{\Delta}(\varphi,p)|l >< s|\hat{\Delta}(\varphi,p)|q >= \delta_{r,q}\delta_{l,s}. \quad (C.4)$$

In fact, substituting (10.28) into the left-hand side of (C.4), and using (10.30) and then (C.2) leads to

$$\sum_p \frac{1}{2\pi}\int d\varphi < r|\hat{\Delta}(\varphi,p)|l >< s|\hat{\Delta}(\varphi,p)|q > \quad (C.5)$$

$$= \sum_m \frac{1}{2\pi}\int d\xi < r|e^{i(\theta m+\hat{n}\xi)}|l >< s|e^{-i(\theta m+\hat{n}\xi)}|q >$$

$$= \sum_m \delta_{r,q} < r|l+m >< s|q-m >= \delta_{r,q}\delta_{l,s}.$$

Substituting (C.4) into (C.3) gives

$$< s|\mathbf{B}|q >=< s|\mathbf{A}|q > .$$

2. Now we prove formula (10.31). Substituting in (10.31) the expressions $a(\varphi,p)$ and $b(\varphi,p)$ (10.27) implies

$$\sum_p \frac{1}{2\pi}\int d\varphi a(\varphi,p)b(\varphi,p) = \sum_p \frac{1}{2\pi}\int d\varphi Tr[\mathbf{A}\hat{\Delta}(\varphi,p)Tr[\mathbf{B}\hat{\Delta}(\varphi,p)] \quad (C.6)$$

$$= \sum_p \frac{1}{2\pi}\int d\varphi \sum_{r,l}\sum_{s,q} < l|\mathbf{A}|r >< r|\hat{\Delta}(\varphi,p)|l >$$

$$< q|\mathbf{B}|s >< s|\hat{\Delta}(\varphi,p)|q > .$$

Using formula (C.4), we obtain from (C.6)

$$\sum_p \frac{1}{2\pi}\int d\varphi a(\varphi,p)b(\varphi,p) = \sum_{r,l}\sum_{s,q} \delta_{r,q}\delta_{l,s} < l|\mathbf{A}|r >< q|\mathbf{B}|s > \quad (C.7)$$

$$= Tr[\mathbf{AB}].$$

3. Here we derive the formula for the evolution of the Wigner function (10.42), (10.43) $\rho_t(\varphi, p)$ for the quantum kicked rotator. From (10.39), (10.26) we have

$$\rho_{t+1}(\varphi, p) = Tr[\hat{U}\hat{\rho}_t\hat{U}^+\hat{\Delta}(\varphi, p)]. \qquad (C.8)$$

Denote by $2\pi\mathbf{G}_Q(\varphi, p|\varphi', p')$ the image of the operator $\hat{U}^+\hat{\Delta}(\varphi, p)\hat{U}$. Then, using (10.31) we derive from (C.8)

$$\rho_{t+1}(\varphi, p) = \sum_p \int d\varphi \mathbf{G}_Q(\varphi, p|\varphi', p')\rho(\varphi', p'), \qquad (C.9)$$

where

$$\mathbf{G}_Q(\varphi, p|\varphi', p') = \frac{1}{2\pi} Tr[\hat{U}^+\hat{\Delta}(\varphi, p)\hat{U}\hat{\Delta}(\varphi', p')] \qquad (C.10)$$

$$= \frac{1}{2\pi}\sum_s < s|\hat{U}^+\hat{\Delta}(\varphi, p)\hat{U}\hat{\Delta}(\varphi', p')|s > .$$

The basis functions $|s>$ in (C.10) have the form (C.3). The following intermediate formulae can easily be proved [6]

$$\hat{U}^+ e^{-i(m\theta+\xi\hat{n})}\hat{U} = e^{i\xi m/2}\sum_k A_k(\xi, m)e^{i(k-m)}e^{(\xi+2\pi\zeta m)\hat{n}}, \qquad (C.11)$$

where $A_k(\xi, m)$ are determined by the formula

$$A_k(\xi, m) = exp(i\pi\zeta m^2)\frac{1}{2\pi}\int d\theta exp\{i\kappa[f(\theta) - f(\theta - 2\pi\zeta m - \xi)] - ik\theta\}. \qquad (C.12)$$

Substituting (10.28), (C.11) in (C.10) and using the definition $A_k(\xi, m)$ (C.12) yields

$$\mathbf{G}_Q(\varphi, p|\varphi', p') = \frac{1}{(2\pi)^2}\sum_m \int_{-\pi}^{\pi} d\xi exp\{i\varphi - \varphi' - 2\pi\zeta p')m\} \qquad (C.13)$$

$$exp\{i(p - p')\xi\}$$

$$\times \sum_k exp(-i\pi\zeta m^2)A_k(\xi, m) exp\left\{i\left(\frac{\xi + 2\pi\zeta m}{2} + \varphi'\right)k\right\}$$

$$= \frac{1}{(2\pi)^2}\sum_m \int_{-\pi}^{\pi} d\xi exp\{i(\varphi - \varphi' - 2\pi\zeta p')m\} exp\{i(p - p')\xi\}$$

$$\times exp\left\{i\kappa\left[f\left(\varphi' + \frac{\xi + 2\pi\zeta m}{2}\right) - f\left(\varphi' - \frac{\xi + 2\pi\zeta m}{2}\right)\right]\right\}.$$

The operator $\mathbf{G}_Q(\varphi, p|\varphi', p')$ possesses the property

$$\sum_p d\varphi \mathbf{G}_Q(\varphi, p|\varphi', p') = 1, \qquad (C.14)$$

therefore, the following condition is satisfied

$$\sum_p \int d\varphi \rho_t(\varphi, p) = \sum_p \int d\varphi \rho_0(\varphi, p) = 1. \qquad (C.15)$$

For $f(\theta) = \cos 2\theta$, we have

$$exp\{i\kappa[\cos(2\varphi' + 2\pi\zeta m + \xi) - \cos(2\varphi' - 2\pi\zeta m - \xi)]\} \qquad (C.16)$$

$$= \sum_l J_l(2\kappa \sin 2\varphi' exp[-i(2\pi\zeta m + \xi)l].$$

Substituting (C.16)in (C.13) yields formula (10.45).

References

1. G.P. Berman, G.M. Zaslavsky, Physica A, 91 (1978) 450.

2. G.P. Berman, A.M. Iomin, G.M. Zaslavsky, Physica D, 4 (1981) 113.

3. M.V. Berry, N.L. Balazs, J.Phys. A, 12 (1979) 625.

4. G. M. Zaslavsky, Phys. Rep., 80 (1981) 157.

5. B.V. Chirikov, F.M. Izrailev and D.L. Shepelyansky, Soviet Scient. Rev., 2C (1981) 208.

6. G.P. Berman and A.R. Kolovsky, Physica D, 8 (1983) 117.

7. H. Frahm and H.J. Mikeska, Z. Phys. B. Cond. Matter, 60 (1985) 117.

8. F. Haake, M. Kus and R. Scharf, Z. Phys. B. Condensed Matter, 66 (1987) 381.

9. M. Toda, K. Ikeda, Phys. Lett. A, 124 (1987) 165.

10. M.E. Goddin, P.W. Milonni, Phys. Rev. A, 37 (1988) 796.

11. K. Nakamura, A.R. Bishop, A. Shudo, Phys. Rev. B, 39 (1989) 12422.

12. B.V. Chirikov, Chaos, 1 (1991) 95.

13. M. Gutzwiller, Chaos in Classical and Quantum Mechanics (Springer-Verlag, New York, 1990).

14. Les. Houches, Section LII, Nato Advanced Study Institute, Chaos and Quantum Physics, eds. M.-J. Giannoni, A. Voros, J. Zinn-Justin, North-Holland, 1991.

15. G.P. Berman, A.R. Kolovsky, Sov. Phys. Usp., 35 (1992) 303.

16. J.E. Bayfield, P.M. Koch, Phys. Rev. Lett., 33 (1974) 258.

17. P.M. Koch, Rydberg States of Atoms and Molecules, ed. by R.F. Stebling and F.B. Dunning (Cambridge University Press, N. Y.), 1983.

18. G. Casati, B.V. Chirikov, D.L. Shepelyansky, I. Guarnery, Phys. Rep., 154 (1987) 77.

19. P.W. Milonni, B. Sundaram, Atoms in Strong Fields: Photoionization and Chaos, E.Wolf, Progress in Optics XXXI, Elsevier Science Publishers B.V., 1993.

20. G.P. Berman, E.N. Bulgakov, G.M. Zaslavsky, Chaos, 2 (1992) 257.

21. G.P. Berman, G.M. Zaslavsky, Logarithm Breaking Time in Quantum Chaos, Preprint LA-UR-92-2521, Los Alamos National Laboratory, U.S.A., 1992.

22. G.P. Berman, E.N. Bulgakov, D.D. Holm, Quantum Chaos of Atoms in a Resonant Cavity at the Influence of an External Resonant Field, Phys. Rev. A, 1994, (in press).

23. G.P. Berman, E.N. Bulgakov, D.D. Holm, V.I. Tsifrinovich, Phys. Lett. A, 181 (1993) 296.

24. R.H. Dicke, Phys. Rev., 93 (1956) 99.

25. S. Haroche, J.M. Raimond, Advances in Atomic and Molecular Physics, 20 (1985) 347.

26. J.M. Raimond, P. Goy, M.Gross, C. Fabre, S. Haroche, Phys.

Rev. Lett., 49 (1982) 1924.

27. Y. Kaluzny, P. Goy, M.Gross, J.M. Raimond, S. Haroche, Phys. Rev. Lett., 51 (1983) 1175.

28. K.N. Alekseev, G.P. Berman, Sov. Phys. JETP, 61 (1985) 569.

29. K.N. Alekseev, G.P. Berman, Phys. Lett. A, 111 (1985) 326.

30. R.J. Glauber, Phys. Rev., 131 (1963) 2766.

31. G.S. Agarval, E. Wolf, Phys. Rev. D, 2 (1970) 2161; 2187; 2206.

32. P. Caruthers, M. Nieto, Rev. Mod. Phys., 40 (1968) 411.

33. W.H. Luisell, Radiation and Noise in Quantum Electronics (New York: McGraw-Hill, 1964).

34. W.M. Zwang, Rev. Mod. Phys., 62 (1990) 867.

35. A.M. Perelomov, Sov. Phys. Usp., 20 (1977) 703.

36. J.M. Radcliffe, J. Phys. A, 4 (1971) 313.

37. F.T. Arecchi, E. Courtens, R. Gilmore, H. Thomas, Phys. Rev. A, 6 (1972) 2211.

38. M. Born, Atomic Physics, Dover Publications, New York, 1992.

39. M. Gross, C. Fabre, P. Pillet, S. Haroche, Phys. Rev. Lett., 36 (1976) 1035.

40. Yu. A. Sinitsin, V.M. Tsukernik, Phys.Lett. A, 90 (1982) 339.

41. O.B. Zaslavsky, Ukr. Fiz. Zh., 29 (1984) 419.

42. R. Courant, D. Hilbert, Methods of Mathematical Physics, Vols. 1,2, Wiley (Interscience), New York, 1966.

43. A. Einstein, in Scientific Papers Presented to Max Born, Hafner Publising Company Inc., NY, 1953, p.33.

44. The Born-Einstein Letters, Walker and Company, NY, 1971, p. 205.

45. M. Born, Continuity, Determinism, and Reality, Dan. Mat. Fys. Medd., 30 (1955) 1.

46. B.V. Chirikov, Phys. Rep., 52 (1979) 263.

47. G.M. Zaslavsky, Chaos in Dynamic Systems (Harwood Academic Publ., NY, 1985).

48. A. Lichtenberg, M. Lieberman, Regular and Chaotic Motion (Springer, 1983).

49. B.V. Chirikov, The Problem of Quantum Chaos, Preprint, Budker Institute of Nuclear Physics, 630090, Russia, 1992.

50. M. Berry, in Chaotic Behavior of Deterministic Systems, edited by R. Helleman and J. Joos (Institute of Theoretical Physics, North-Holland, Amsterdam, 1983),pp. 173-271.

51. B. Eckhardt, Phys. Rep., 163 (1988) 205.

52. P.V. Elutin, Sov. Phys. Usp., 31 (1988) 597.

53. L.E. Reichl, The Transition to Chaos in Conservative Classical Systems: Classical Manifistations (Springer-Verlag), 1992.

54. F. Haake, Quantum Signature of Chaos, (Springer-Verlag), 1990.

55. G.M. Zaslavsky, M.Yu. Zakharov, R.Z. Sagdeev, D.A. Usikov, A.A. Chernikov, Zh. Eksp. Teor. Fiz., 91 (1986) 500.

56. A.A. Chernikov, R.Z. Sagdeev, D.A. Usikov, M.Yu. Zakharov, G.M. Zaslavsky, Nature, 326 (1987) 559.

57. B.V. Chirikov, Phys. Rep., 52 (1978) 263.

58. S.R. Channon, J.L. Lebowitz, Ann. NY Acad. Sci., 357 (1980) 108.

59. B.V. Chirikov, D.L. Shepelyansky, Proc. 9-th Int. Conf. on Nonlinear Oscillations (Kiev, 1981), Vol. 2 (Moscow: Naukova Dumka),p. 421, (Engl. transl. 1983 PPL-TRANS-133, Plasma Physics Lab., Princeton University).

60. C.F.F.Karney, Physica D, 8 (1983) 360.

61. B.V. Chirikov, D.L. Shepelyansky, Chaos Border and Statistical Anomalies, Preprint 86-174, Inst. of Nuclear Physics, Novosibirsk, Russia, 1986.

62. G.P. Berman, G.M. Zaslavsky, Physica A, 97 (1979) 367.

63. G.P. Berman, G.M. Zaslavsky, Phys. Lett. A, 61 (1977) 295.

64. G.P. Berman, G.M. Zaslavsky, A.R. Kolovsky, Sov. Phys. JETP, 54 (1981) 272.

65. G.P. Berman, G.M. Zaslavsky, A.R. Kolovsky, Phys. Lett. A, 87 (1982) 152.

66. G.P. Berman, G.M. Zaslavsky, A.R. Kolovsky, Sov. Phys. JETP, 61 (1985) 925.

67. D.F. Escande, Phys. Rep., 121 (1985) 165.

68. G.P. Berman, V.Yu. Rubaev, G.M. Zaslavsky, Nonlinearity, 4 (1991) 543.

69. G.M. Zaslavsky, R.Z. Sagdeev, D.A. Usikov, A.A. Chernikov, Sov. Phys. Usp., 30 (1987) 436.

70. A.J. Lichtenberg, B.P. Wood, Phys. Rev. A, 39 (1989) 2153.

71. G. Casati, B.V. Chirikov, J. Ford, F.M. Izrailev, Stochastic Behavior in Classical and Quantum Hamiltonian Systems (Springer Lecture Notes in Physics,), ed. G. Casati and J. Ford, 93 (1979) 334.

72. M. Hillery, R.F. O'Connell, M.O. Scully, E.P. Wigner, Phys. Rep., 106 (1984) 121.

73. V.I Tatarsky, Usp. Fiz. Nauk, 139 (1983) 587.

74. E.P. Wigner, Phys. Rev., 40 (1932) 749.

75. M.V. Berry, N.L. Balazs, M. Tabor, A. Voros, Ann. Phys., 122 (1979) 26.

76. H.Y. Korsch, M.V. Berry, Physica D, 3 (1981) 627.

77. M.B. Berry, Phil. Trans. Roy. Soc. A, 287 (1977) 237.

78. N. Makunda, An. J. Phys., 47 (1979) 182.

79. S. Fishman, D.R. Grempel, R.E. Prange, Phys. Rev. Lett., 49 (1982) 509.

80. R. Blümer, S. Fishman, M. Grimasti, Y. Smilansky, Lecture Notes in Physics,(Berlin: Springer), 263 (1986) 212.

81. B.V. Chiricov, F.M. Izrailev, D.L. Shepelyansky, Physica D, 33 (1988) 77.

82. E.R. Prange, D.R. Crempel, S. Fishman, Phys. Rev. B, 29 (1984) 6500.

83. D.L. Shepelyansky, Physica D, 8 (1983) 208.

84. S. Adachi, M. Toda, K. Ikeda, Phys. Rev. Lett., 61 (1988) 659.

85. F.M. Izrailev, D.L. Shepelyansky, Teor. Mat. Fiz., 43 (1980) 417.

86. G.P. Berman, A.R. Kolovsky, Physica D, 17 (1985) 183.

87. D.L. Shepelyansky, Teor. Mat. Fiz., 49 (1981) 117.

88. S. Fishman, D.R. Crempel, R.E. Prange, Phys. Rev. A, 29 (1984) 639.

89. B. Dorizzi, B. Grammaticos, Y. Pomeau, J. Stat. Phys., 37 (1984) 93.

90. G.P. Berman, A.R. Kolovsky, F.M. Izrailev, Physica A, 152 (1988) 286.

91. G.P. Berman, A.R. Kolovsky, F.M. Izrailev, A.M. Iomin, Chaos, 1 (1991) 220.

92. R. Graham, Physica Scripta, 35 (1987) 111.

93. S. Stenholm, Phy, Rep., 6 (1073) 3.

94. P.I. Belobrov, G.M. Zaslavsky, G.Kh. Tartakovsky, Sov. Phys.

JETP, 44 (1976) 945.

95. R. Graham, M. Hohnerback, Z. Phys. B, 57 (1984) 233.

96. R. Graham, M. Hohnerback, Phys. Lett. A, 101 (1984) 61.

97. R.F. Fox, J. Eidson, Phys. Rev. A, 34 (1986) 482.

98. R.F. Fox, J. Eidson, Phys. Rev. A, 36 (1987) 4321.

99. R.F. Fox, in The Ubiquity of Chaos, edited by S. Krasner (American Association of the Advancement of Science, Washington, Dc 1990), p. 105.

100. N. Klenner, J. Weis, M. Doucha, J. Phys. C, 19 (1986) 4673.

101. M.B. Cibils, Y. Cuche, W.F. Wreszinski, J.-P. Amiet, H. Beck, J. Phys. A: Gen., 23 (1990) 545.

102. K.N. Alekseev, G.P. Berman, Sov. Phys JETP, 65 (1987) 1115.

103. D. D. Holm, G. Kovačič, B. Sundaram, Phys. Lett. A, 154 (1991) 346.

104. D. D. Holm, G. Kovačič, Physica D, 56 (1992) 270.

105. R. Bonifacio, G. Preparata, Phys. Rev. A, 2 (1970) 336.

106. A.N. Holden, C. Kittel, F.R. Yager, Phys. Rev., 77 (1950) 147.

107. V.W. Cohen, C. Kikuchi, J. Turkevich, Phys. Rev., Phys. Rev., 85 (1952) 379.

108. S.A. Altshuller, B.M. Kozirev, Electronic Paramagnetic Resonance, M., Fizmatgiz, 1961.

109. M. Gell-Mann, J.B. Hartle, in Proceedings of the 3rd International Symposium Foundations of Quantum Mechanics in the Light of New Technology, edited by S. Kobayashi, H.Ezawa, Y. Murayama, and S. Nomura (Physical Society of Japan, Tokyo), 1990.

110. R. Omnes, Rev. Mod. Phys., 64 (1992) 339.

111. A.N. Gorban, V.A. Okhonin, Universality Domain for the Statistics of Energy Spectrum, Preprint KIP-29, Krasnoyarsk, 1983.

112. C.L. Mehta, E.C.G. Sudarshan, Phys.Lett., 22 (1966) 574.

113. L. Mista, Phys. Lett. A, 25 (1967) 646.

114. Y. Kano, Phys. Lett. A, 56 (1976) 7.

115. A. Trias, Phys. Lett. A, 61 (1977) 149.

116. G.P. Berman, G.M. Zaslavsky, Physica A, 111 (1982) 17.

117. G.M. Zaslavsky, Phys. Lett. A, 69 (1978) 145.

118. L.G. Yaffe, Rev. Mod. Phys., 54 (1982) 407.